Springer Proceedings in Business and Economics

Springer Proceedings in Business and Economics brings the most current research presented at conferences and workshops to a global readership. The series features volumes (in electronic and print formats) of selected contributions from conferences in all areas of economics, business, management, and finance. In addition to an overall evaluation by the publisher of the topical interest, scientific quality, and timeliness of each volume, each contribution is refereed to standards comparable to those of leading journals, resulting in authoritative contributions to the respective fields. Springer's production and distribution infrastructure ensures rapid publication and wide circulation of the latest developments in the most compelling and promising areas of research today.

The editorial development of volumes may be managed using Springer's innovative Online Conference Service (OCS), a proven online manuscript management and review system. This system is designed to ensure an efficient timeline for your publication, making Springer Proceedings in Business and Economics the premier series to publish your workshop or conference volume.

More information about this series at http://www.springer.com/series/11960

Luigi Paganetto
Editor

Capitalism, Global Change and Sustainable Development

Springer

Editor
Luigi Paganetto
FUET – Tor Vergata Economics Foundation
University of Rome Tor Vergata
Rome, Italy

ISSN 2198-7246 ISSN 2198-7254 (electronic)
Springer Proceedings in Business and Economics
ISBN 978-3-030-46142-3 ISBN 978-3-030-46143-0 (eBook)
https://doi.org/10.1007/978-3-030-46143-0

This Springer imprint is published by the registered company Springer Nature Switzerland AG.
The registered company address is: Gewerbestrasse 11, 6330 Cham, Switzerland

Contents

Slower World Growth or Global Recession with the Trade War?

Dominick Salvatore

Abstract After the slow recovery from the 2008–2009 global financial crisis, the world economy now faces even slower growth and the prospects of global recession triggered by the USA–China trade war. This paper starts by examining whether the slow growth of the world economy was due to wrong or inadequate economic policies to overcome the crisis (great recession) and return to growth, or for other reasons. Then, it examines the reasons the United States is growing faster than other advanced countries, the slowing growth of emerging market economies (and even economic crisis in some of them), and whether the world is now sliding toward a new global financial crisis and recession triggered by the current trade war.

Keywords Growth prospects · Advanced countries · Emerging market economies · Global financial crisis · Trade war · New global financial crisis

1 Introduction

After a decade from the end of the deepest global financial crisis of the postwar period, growth continues to be slow in advanced countries and falling in most emerging market economies. There is now even the risk that the world could be drifting toward a new global financial crisis triggered by the US–China trade war.

This paper examines the reasons for (1) the slow recovery after the recent global financial crisis, (2) the United States growing faster than other advanced countries (particularly Europe), (3) growth slowing in emerging market economies (with some already in crisis), and (4) the prospects of the world facing a new global financial and economic crisis—possibly even deeper than the previous crisis.

D. Salvatore (✉)
Department of Economics, Fordham University, New York, USA
e-mail: salvatore@fordham.edu

L. Paganetto (ed.), *Capitalism, Global Change and Sustainable Development*, Springer Proceedings in Business and Economics,
https://doi.org/10.1007/978-3-030-46143-0_1

2 Causes and Effects of the "Great Recession"

The most recent global financial crisis started in the U.S. housing sector in 2007 as a result of banks giving huge amounts of (sub-prime) loans or mortgages to individuals and families that could not afford them. When many individuals and families defaulted on their loans, U.S. banks fell into a deep crisis, which then spread to the entire financial sector in 2008 and, from there, to the U.S. real sector and the rest of the world. The result was the "Great Recession".

Contagion spread from the United States across the Atlantic because many European banks had committed even greater excesses than U.S. banks and Europe faced an even greater housing bubble than the United States (Salvatore 2010). Deep recession in all advanced countries then greatly reduced their imports from and financial flows to emerging market economies, thereby spreading the crisis to the rest of the world. Most emerging market economies (such as Argentina, Brazil, Mexico, Russia, and Turkey) fell into a deep recession, while others (such as China) faced a sharp growth slowdown.

Table 1 shows that at the depth of the recession in 2009, the percentage fall of real GDP among the largest advanced nation ranged from 2.8 for the United States to 5.6 for Germany, and it was 4.5 in the Euro Area as a whole, 2.9 in Canada, 4.3 in the United Kingdom, and 5.4 in Japan. The recovery from 2010 to 2017 was slow in all advanced nations, averaging 2.1% in the United States, 1.3% in the EU-19 or Euro Area (which even fell back into recession in 2012 and 2013), 2.0% in the United Kingdom, 1.5% in Japan, and 2.3% in Canada. Britain and Canada fared generally better than the Euro Area and Japan since the end of the crisis, with growth rates comparable to those in the United States. Japan suffered 6 recessions during the past 25 years (including in the ones 2011 and 2014).

In 2018, growth accelerated somewhat in the EU-19 and in some countries (United States, France, Italy and Spain) but declined in others (Germany, United Kingdom, Japan, and Canada) and it is forecasted to decline in 2019 in the EU-19

Table 1 Average annual percentage growth of real GDP in major advanced countries in 2009–2018 and forecasts for 2019–2020

Nation/area	Yearly average				
	2009	2010–2017	2018	2019[a]	2020[a]
USA	−2.8	2.1	2.9	2.3	1.9
EU-19	−4.5	1.3	1.8	1.3	1.5
Germany	−5.6	2.1	1.5	0.8	1.4
France	−2.9	1.3	1.5	1.3	1.4
Italy	−5.5	0.2	0.9	0.1	0.9
Spain	−3.6	0.7	2.5	2.1	1.9
U.K.	−4.3	2.0	1.4	1.2	1.4
Japan	−5.4	1.5	0.8	1.0	0.5
Canada	−2.9	2.3	1.8	1.5	1.9

[a]Estimates
Source: IMF (2019a)

and in all the countries in Table 1 (except Japan), but to be higher than in 2019 in 2020 (except in the United States, Spain and Japan).

3 Countercyclical Monetary and Fiscal Policies to Overcome the Crisis and Restart Growth

The United States and other advanced nations responded to the Great Recession by rescuing banks and other financial institutions from bankruptcy, slashing interest rates, introducing huge economic stimulus packages, and undertaking huge injections of liquidity (quantitative easing or QE). These efforts, however, only succeeded in preventing the economic recession from being deeper than otherwise and the subsequent recovery to be even slower than it has been.

In order to overcome the crisis, the Fed, the European Central Bank (ECB), and the Bank of Japan (as well as the central Bank of England and Canada) drastically cut their fund rate starting in 2008, so much so that by early 2014 these rates were close to zero in nominal terms and negative in real terms. These nations also conducted quantitative easing (QE) that sharply increased their holdings of private long-term assets so as to further lower long-term rates in the hope of stimulating growth (see Fig. 1). More recently, the ECB and the Bank of Japan (as well the central bank of Denmark, Sweden, and Switzerland) even introduced negative *nominal* interest rate. Such powerful expansionary policy had been not envisioned until the recent crisis.

Advanced nations also used powerful expansionary fiscal policy to fight the recession and restart growth, which led to huge budget deficits and government debts. Table 2 shows that the U.S. budget deficit as a percentage of GDP was 2.9 in 2007, reached the high of 13.1 in 2009, before falling to 3.2 in 2015, and then rising to 3.8 in 2017 and 4.3 in 2018. In the Euro Area, comparable figures were 0.6, 6.3,

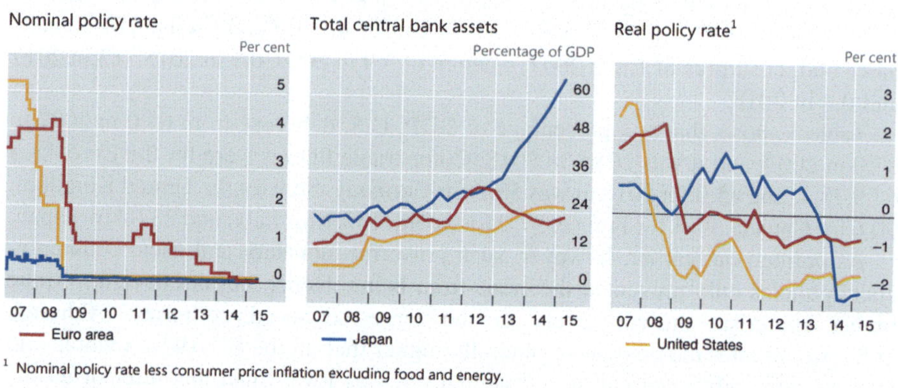

Fig. 1 Nominal and real policy rates, and holdings of Total Central Bank Assets: Euro Area, Japan and United States, 2007–2015 (Source: BIS 2015)

Table 2 Budget deficits as percentage of GDP in major advanced countries, 2007–2019

Nation/area	2007	2009	2011	2013	2015	2017	2018	2019[a]
United States	−2.9	−13.1	−9.3	−4.1	−3.2	−3.8	−4.3	−4.6
Euro Area	−0.6	−6.3	−4.2	−3.1	−2.0	−1.0	−0.6	−1.0
Germany	0.3	−3.0	−1.0	−0.1	0.8	1.0	1.7	1.1
France	−2.5	−7.2	−5.2	−4.1	−3.6	−2.7	−2.6	−3.3
Italy	−1.5	−5.3	−3.7	−2.9	−2.6	−2.4	−2.1	−2.7
Spain	2.0	−11.0	−9.6	−7.0	−5.3	−3.1	−2.7	−2.3
United Kingdom	−3.0	−10.7	−7.5	−5.3	−4.2	−1.8	−1.4	−1.3
Japan	−2.1	−10.4	−9.4	−7.9	−3.8	−3.2	−3.2	−2.8
Canada	1.8	−3.9	−3.3	−1.5	−0.1	−0.3	−0.4	−0.6

[a]Estimate
Source: IMF (2019b)

Table 3 Government debt as percentage of GDP in major advanced countries, 2007–2019

Nation/area	2007	2009	2011	2013	2015	2017	2018	2019[a]
United States	64.0	87.0	99.7	104.8	104.7	106.2	105.8	106.7
Eurozone	64.9	78.4	86.6	91.6	89.9	86.8	85.0	83.6
Germany	63.6	72.6	78.6	77.4	70.8	63.9	59.8	56.9
France	64.4	82.9	87.8	93.4	95.6	98.5	98.6	99.2
Italy	99.8	112.5	116.5	129.0	131.6	131.3	132.1	133.4
Spain	35.5	52.7	69.5	95.5	99.3	98.1	97.0	96.0
United Kingdom	42.2	64.1	80.8	85.2	87.9	87.1	86.9	85.7
Japan	183.0	201.0	222.1	232.5	231.6	235.0	237.2	237.5
Canada	66.8	79.3	81.9	86.2	91.3	90.1	90.6	88.0

[a]Estimate
Source: IMF (2019b)

2.0, 1.0, and 0.6. In the United Kingdom, it was 3.0 in 2007, reached the high of 10.7 in 2009, but then fell to 4.2 in 2015, 1.8 in 2017 and 1.4 in 2018. In Japan it was 2.1 in 2007, reached 10.4 in 2009, with 3.2 in 2015, 3.2 in 2017 and 1.4 in 2018. Canada had a budget surplus of 1.8 in 2007, a deficit of 3.9 in 2009, 0.1 in 2015, 0.3 in 2017 and 0.4 in 2018.

Table 3 shows that as a percentage of GDP, U.S. debt rose from 64.0 in 2007 to 87.0 in 2009, and it was 105.8 in 2018. Comparable figures were for the Euro Area 64.9,78.4, and 85.0. In 2018, it was 59.8 for Germany, 86.9 for the United Kingdom, 90.6 for Canada, 96.8 for France, 97.0 for Spain, 132.1 for Italy, and 237.2 for Japan.

As pointed out earlier, however, such powerful monetary and fiscal countercyclical policies only prevented the crisis from being deeper, but not to return to the higher pre-crisis growth levels in the large advanced countries. Although U.S. growth after the crisis was generally higher than in the EU-19 as a whole and of most large advanced nation individually, it was lower than that after of all the previous postwar recessions (see Salvatore 2018).

4 Slow Recovery in United States but Faster Than Europe's

Although the U.S. recovery after the recent crisis was the slowest than after all its post-war recessions (see Salvatore 2016), it was faster than in Europe (see Fig. 2). The figures shows, however, that Europe's growth problem started much earlier (in the 1990's) than with the recent crisis, but it became even worse as a result with the crisis (i.e., the diverge from U.S. growth became even worse after the crisis).

But why was the Great Recession so deep and the recovery so slow in the United States and other advanced countries? One important reason is that the Great Recession was triggered and accompanied by a banking and financial crisis, and experience indicates (Reinhart and Rogoff 2009) that this type of crisis is much more difficult and usually takes much longer to overcome than a purely economic crisis because it is usually accompanied by heavy deleveraging by the banking sector. But there are also other more specific reasons that made this recovery and growth so slow.

The most important of these is that advanced nations faced a sharp decline in labor productivity after the recent global financial crisis. Table 4 shows that while U.S. labor productivity increased at a yearly average of 2.4% in 1999–2006, it rose by only 1.1% in 2007–2015. Comparable figures (in percentages) were, respectively, 1.8 and 0.7 for Japan, 2.4 and 0.1 in the United Kingdom, 1.5 and 0.5 for the Euro Area, and 1.9 and 0.6 labor for the European Union of 28 (EU-28).

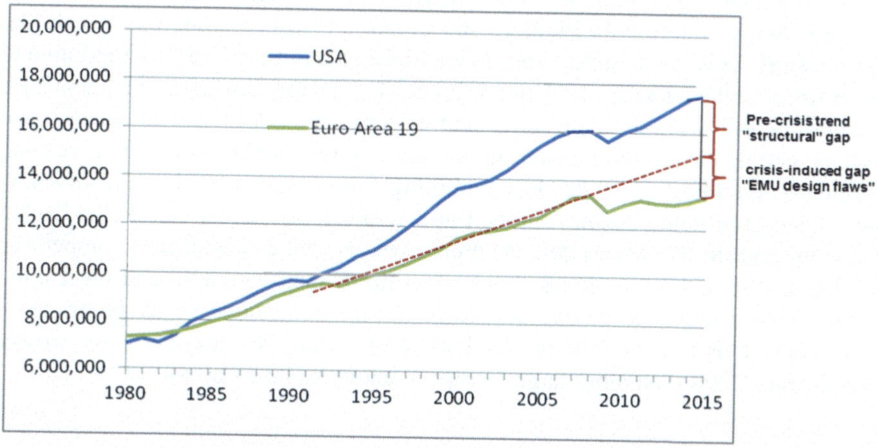

Fig. 2 The Euro Area growth crisis goes far back, but became worse after the global financial crisis (Source: Conference Board 2016)

Table 4 Growth of labor productivity, average percent per year, 1999–2015

	US	Japan	UK	Euro Area	EU-28
1999–2006	2.4	1.8	2.4	1.5	1.9
2007–2015	1.1	0.7	0.1	0.5	0.6

Source: OECD (2016)

Another reason for slower growth (and less job creation) in the United States and other advanced countries was outsourcing. Most multinationals from advanced nations, in their effort to minimize production costs, transferred a great deal of production (and jobs) to emerging market economies, especially China, during the past decade. This started before the recent global financial crisis but it rapidly accelerated afterwards. Growth was also discouraged in advanced nations by high taxes, overregulation and policy uncertainty—again, more so in Europe that in the United States and the United kingdom (as shown in Fig. 2).

5 Why Is Growth Slower in Europe Than in the United States?

Overregulation and excessive welfare benefits are generally regarded as the primary reasons for the slower growth in Europe than in the United States (Salvatore 1998, 2007, 2017). Europeans expect their government to take care of them from birth to death. They have guaranteed vacations, free health care, free education, practically guaranteed jobs, and high unemployment benefits. All this kept the European growth rate much lower than growth in the United States during the past two decades and landed Europe into a deeper recession and a weaker recovery than the United Stares (Salvatore 1998, 2004). Japan is in a similar situation.

Overregulation and high taxes lower efficiency, increases the cost of doing business, and slows growth. Figure 3 shows that the cost of doing business in the Euro Area (EA) is about almost four times higher than in the United Kingdom and seven times higher than in the United States. The Doing Business (DB) Indicator includes the following ten indicators: starting a business, dealing with construction permits, getting electricity, registering property, getting credit, protecting investors, paying taxes, trading across borders, enforcing contracts, and resolving insolvency. Since the cost of doing business in the Euro Area is much higher than in the United Kingdom and the United States, the Euro Area is less efficient, less competitive internationally, and grows more slowly than the United Kingdom and the United States. Italy stands out among the large Euro Area countries as being more overregulated and having higher welfare costs—and thus experiencing slower growth since 2008 than other large European countries (see Fig. 4).

6 Growth Slowdown or Recession in Major Emerging Market Economies

After their spectacular growth during the past decade, growth also slowed down in emerging market economies (Klein and Salvatore 2013). The economic crisis in emerging market economies started as a result of contagion as the recession-afflicted

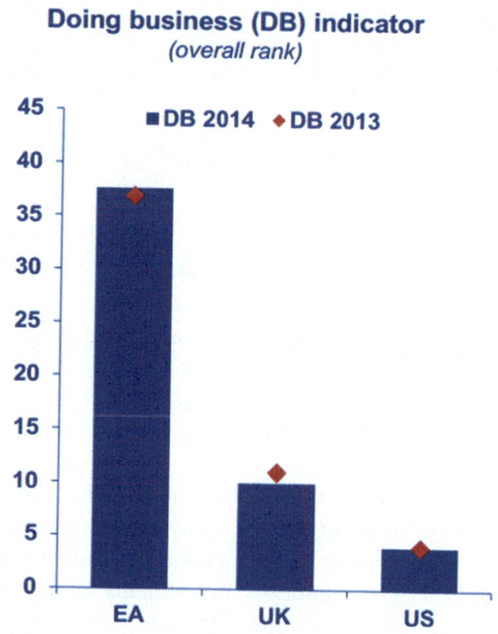

Fig. 3 Doing business indicator (Source: World Bank 2016)

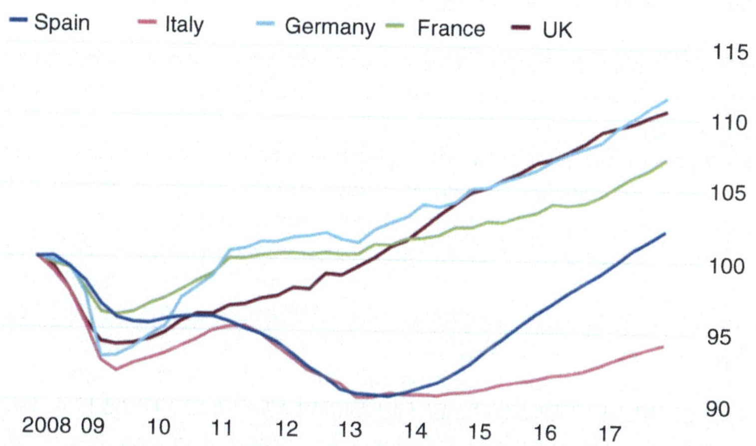

Fig. 4 Real GDP Growth in Italy, Spain, France, Germany and U.K, 2008–2017 (Source: OECD 2018)

advanced nations sharply cut imports from and the flow of investments to emerging markets during the recent global financial crisis. This was one reason for the decline in China's previous spectacular growth. While India's growth held up, the average annual percentage growth of real GDP in other large emerging countries (those in the G20) has been very slow (with exception of China, India and Indonesia—see Table 5).

Table 5 Average annual percentage growth of real GDP in emerging countries in G20 in 2009–2018 and forecast for 2019–2020

Nation/area	2009	2010–2017	2018	2019[a]	2020[a]
China	9.2	8.0	6.6	6.3	6.1
India	8.5	7.3	7.1	7.3	7.5
Russia	−7.8	1.9	2.3	1.6	1.7
Brazil	−0.1	0.4	1.1	2.1	2.5
S. Africa	−1.5	2.0	0.8	1.2	1.5
Korea	0.7	3.5	2.7	2.6	2.8
Indonesia	4.7	5.5	5.2	5.2	5.2
Mexico	−4.7	3.2	2.0	1.6	1.9
Argentina	−5.9	2.4	−2.5	−1.2	2.2
Turkey	−4.7	6.8	2.6	−2.5	2.5
S. Arabia	−2.1	4.0	2.2	1.8	2.1

[a]Estimates
Source: IMF (2019a)

The slowdown in China's growth rate from 2010 to 2017, however, was also the result of its effort to restructure its economy toward more internal demand and from a manufacturing to a service economy, rather than relying on continued export growth and on heavy domestic investments as in the past (the former because it was no longer sustainable and the latter because of the setting in of the "dreaded" diminishing returns). In the process, China's demand for primary commodity imports from other emerging markets (especially from Brazil and many African countries) also declined sharply. This caused an even greater growth slowdown in other emerging market economies than the decline resulting from the recession in advanced countries.

As the demand for their primary exports declined, the currencies of emerging markets sharply depreciated or were devalued, thus making the servicing of their previously accumulated heavy financial debt and dollar borrowings of over $6 trillion by 2018 unsustainable. Emerging market economies also experienced large net financial outflows, which exacerbated their economic problems even more. Thus, Brazil faced recession with a decline of real GDP of 3.5 in 2015 and 2016, Russia faced even more economic difficulties as a result also of the wars and the economic sanctions imposed on it, and Argentina and Turkey also faced an economic crisis.

The economic crisis in most emerging market economies would have been even deeper had they not been operating under some form of exchange rate flexibility or had not devalued their currencies (which cushioned to some extent their reduction in exports) and had they not also entered the crisis with more foreign exchange reserves accumulated during the commodity boom of the previous decade.

7 The U.S. Trade War with China

The United States started a trade war with China in order to force China to end its unfair trade practices. The *White House Office of Trade and Manufacturing Policy* published a study (June 2018) providing evidence on "How China's Economic Aggression Threatens the Technologies and Intellectual Properties of the United States and the World" by engaging in: (1) systematic stealing U.S. technology on a grand scale sponsored by the Chinese government, (2) providing illegal subsidies to some of its large companies to overcome foreign competition, (3) requiring U.S. and other foreign firms to enter into joint ventures and transfer their technology to Chinese companies as a condition to operate in China, and (4) engaging in the over-expansion of its production capacity of many products (such as steel and aluminum) and then dumping the excess production on the U.S. and other world markets.

China was admitted to the World Trade Organization (WTO) in 2002 on the condition that it would gradually move toward a market economy over time and follow WTO's trade rules. That did not occur. China took full advantage of the opportunities of, more or less, free access to other countries' markets but it did not allow equal access to foreign companies in the Chinese market. Clearly, China "gamed the system". Over the years, China promised several times to change its ways, but it never did and does seem willing to do so. This set the stage for President Trump imposing higher and higher import tariffs on more and more Chinese exports to the United States. The purpose was (and is) to try to pry open the Chinese market to U.S. companies, end Chinese steeling of foreign technology, and ending all its other unfair trade practices, so as to establish a level-playing field in international trade and economic relations.

The U.S. accusation that China is not following WTO rules seems to be mostly true and is more or less believed also by the European Union, Japan and Canada. China is now responsible for more than half of the U.S. trade deficit. Of course, there is no theory that postulates that a nation needs to balance trade bilaterally. But when one nation (in this case China) is responsible for more than half of the total trade of the very large and unsustainable trade deficit of the biggest (and very competitive) economy in the world (the United States)—clearly something is wrong.

Of course, Trump was told that import tariffs on Chinese and other nations' exports to the United States would not reduce the U.S. trade deficit because international trade theory postulates that trade deficits are caused (result) from gross *domestic* macro imbalances, specifically from the United States spending much more than it saves, with the excess reflected in its trade deficit. Trump used import tariffs primarily as a weapon to force its trade partners, especially China, to open their markers and reduce their protectionism against the United States to the lower U.S. level.

The United States (Trump) believes that it is fair and appropriate to protect itself by imposing "countervailing duties" on China's exports to the United States, just as it would be prescribed by the WTO rules—and without allowing China to retaliate (as it occurred when the United States imposed a 25% import tariff on $50 billion of Chinese exports to the United States in July–August 2018). However, the United States did take China to the WTO and this allowed China to immediately retaliate. The United States then "doubled up" by imposing a tariff of 10% on another $200 billion of Chinese exports to the United States in September 2018 (raised to 25% in June 2019—and China again immediately retaliated). The stage was thus set for a full-fledged trade war. In fact, the United States has indicated that on October 1, 2019 it will impose a 15% tariff on a subset of the remaining $300 billions of Chinese exports to the United States, and then to the whole $300 billion on December 15. The United States also prohibited its high-tech companies to sell advanced semiconductors and other high-tech products to Chinese firms and from acquiring or invest in high-tech U.S. companies.

The United States also took or threatened protectionist trade action against other trade partners. Under President Trump, the United States dropped out of the Tans Pacific Partnership (TPP), put the Transatlantic Trade and Investment Partnership (TTIP) on hold, forced the renegotiation of NAFTA on Canada and Mexico, imposed hundreds of billion of dollars of tariffs on U.S. imports of steel, aluminum and other products on most its major trade partners, and threatened to impose a 25% tariff on automobile imports from Europe, Japan, China and other nations if they did not reduce their tariffs on U.S. exports to U.S. levels. Trump even threatened to drop out of the World Trade Organization (WTO). Trump acted unilaterally because of his strong belief that the WTO was unable or unwilling to enforce its trade rules. But by so doing, Trump seriously undermined the WTO and is risking unleashing a new global economic crisis.

The correct U.S. trade policy would have been and should be to convince the other major trading nations that China's non-market trading practices harms them all, not just the United States, and convince them to jointly bring China to the WTO, where legal proof of China's transgressions would be presented which, if proved true, would lead the WTO to sanction the joint legal imposition of countervailing duties on China (and without permitting China to retaliate). This policy would have a greater chance of China's accepting to change its ways in international competition and trade, and end the trade war.

8 Will the U.S. Trade War with China Lead to a New Global Crisis and Recession?

Figure 5 provides an OECD simulation scenario that shows the effect of an increase in trade protection by the United States, Europe, and China that increased trade costs by 10%, and the resulting percentage deviation in the GDP, imports, and exports of the United States, China, Europe, and the rest of the world. The simulation exercise shows that GDP, imports, and exports would fall by various percentages with respect to the baseline in all countries and areas or groups of countries, and the world as a

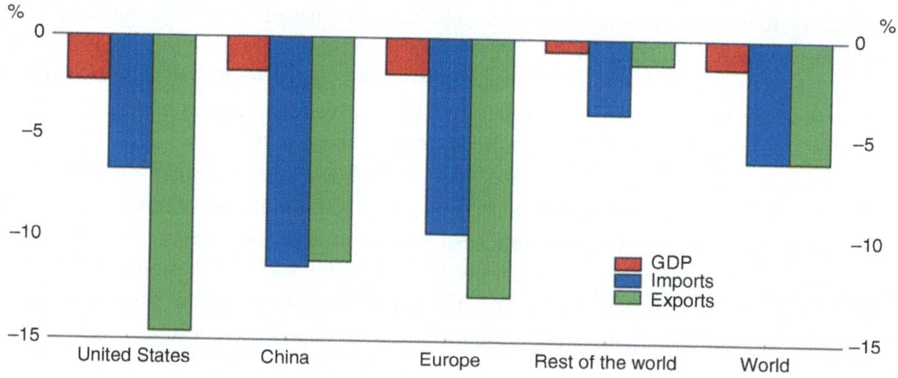

Fig. 5 Effect on GDP, Imports, and Exports of an Increase in Trade Protection by the United States, Europe and China that raises trade costs by 10% (Source: OECD 2016)

whole. The figure shows that the U.S. GDP would fall more than in other nations and areas. Imports would fall the most in China and Europe, while exports would fall more in the United States and Europe (for the result of another similar simulation exercise, see Salvatore and Campano 2018).

Of course, an all-out trade war would very likely be much more economically damaging than indicated in Fig. 5 because the tariffs on U.S.–China trade have increased much more than the 10% assumed in the OECD simulation shown in Fig. 5. In fact, U.S. tariffs on Chinese exports to the United States rose from 3.1% in January 2018 to 24.3% in November 2019 and Chinese retaliatory tariffs on U.S. export increased from 8.0 to 25.9% over the same time period. On the other hand, trade tariffs on the rest of world trade have not increased.

A more recent OECD (2019) estimate indicates that the U.S–China trade restriction imposed so far (from July 2018 through August 2019) would cut 1.0 percentage point of growth in China over the next 2–3 years, 0.7 of growth in the United States (thus lowering its growth from the forecasted rate of 2.3% in 2019 and 1.9% in 2020 to 1.7 and 1.3, respectively), and from and 0.6 growth of the world as a whole. But if the trade war progresses and intensifies (as it seemed likely in fall 2019), the economic harm in terms of the reduction in world growth could be much higher, especially if other nations were drawn into the trade war.

The U.S.–China trade war is almost certain to also worsen other economic weaknesses present in the world economy today. These are: (1) the excessive total indebtedness in the Chinese economy, now in excess of 275% of its GDP, (2) emerging markets' total indebtedness in dollars, which now exceeds $6 trillion and in the face of an appreciating dollar, (3) the new financial bubble arising from nominal interest rates near zero in the United States, Britain and Canada and negative in Japan, the Euro Area (and Denmark, Sweden and Switzerland), thus leading inflated stock markets and also investors, in search for returns, to undertake "excessively" risky investments, and (4) discouraging world consumption and investments.

Each of these weaknesses or dangers could in fact trigger a new global financial crisis and recession on their own even without the trade war, but in the current trade war situation they could lead to "a perfect storm" (i.e., a deeper global financial crisis deeper than the previous one) at a time when most advanced nations have fewer (and less effective) monetary and fiscal policies tools to deal with it.

The United States and China are now (October 2019) engaged in intensive negotiations to avoid an escalation in trade protectionism and worsen the trade war, but even if an accord were reached, strong international trade controversies are likely to recur, especially between the United States and China, unless the WTO were reformed and strengthened, and the world returned to the rule of law in international economic and trade relations and to a liberal economic and trade system.

9 Conclusion

After the generally slow recovery from the recent global financial crisis and great recession, the world is again in danger of sliding into a new crisis triggered by the present U.S.–China trade war, which could be even deeper than the previous crisis because of other current weaknesses in the world economy. The hope is that reason will prevail by ending the trade war and correcting the other weaknesses and dangers currently facing the world economy, not only to avoid another global financial crisis but also to return to the faster world growth that prevailed before the recent crisis.

References

Bank for International Settlements (BIS). (2015). *Annual Report. Basel.*

Conference Board. (2016). *Total economy Database*. Washington, DC.

IMF. (2019a, April). *World economic outlook.* Washington, DC.

IMF. (2019b, April). *Fiscal monitor.* Washington, DC.

Klein, L., & Salvatore, D. (2013). From G-7 to G-20. *Journal of Policy Modeling, May/June*, 416–424.

OECD. (2016, November). *OECD Economic Outlook*. Paris: OECD.

OECD. (2018, May). *OECD Economic Outlook.* Paris: OECD.

OECD. (2019, May). *OECD Economic Outlook.* Paris: OECD.

Reinhart, C. M., & Rogoff, K. (2009). *This time is different.* Princeton University Press: Princeton, NJ.

Salvatore, D. (1998). Europe's structural and competitiveness problems and the Euro. *The World Economy, 21*, 189–205.

Salvatore, D. (2004, October). Globalization, comparative advantage and Europe's double competitive squeeze. *Global Economy Journal, 4*, 1–18.

Salvatore, D. (2007). The U.S. challenge of European firms: Globalization, architecture, and perceived innovativeness. *European Journal of International Management, 1*(1/2), 69–80.

Salvatore, D. (2010). Causes and effects of the global financial crisis. *Journal of Politics and Society, 21*(Spring), 7–16.

Salvatore, D. (2016). Slow recovery and growth for the United States. *Journal of Policy Modeling, 38*(July/August), 624–631.
Salvatore, D. (2017). Europe's growth crisis: When and how will it end? *The World Economy, 40* (May), 836–848.
Salvatore, D. (2018). Trump economics: Overview of effects on the United States and the World. *Journal of Policy Modeling, 40*, 480–488.
Salvatore, D., & Campano, F. (2018). Simulating some of the administration's proposed trade policies. *Journal of Policy Modeling, 40*(May/June), 636–646.
White House Office of Trade and Manufacturing Policy. (2018). *China's economic aggression.* DC: Washington.
World Bank. (2016). *Doing business.* Washington, DC.

European Solidarity as a Way to Face Globalization and as an Antidote Against Populism

Pompeo Della Posta

Abstract Two opposite positions are facing each other in the euro area. The first one, originating in southern Europe, is based on the observation that the adoption of the euro favored northern economies, allowing them to have persistently high trade surpluses at the expenses of the southern ones. The second position, originating in northern Europe, focuses instead on the risks that the citizens of those countries would incur to cover the high debt of southern countries. Such opposed positions, however, ignore the deep reasons that led European countries to start the process of economic integration, of which monetary integration is a relevant part. In view of what precedes, a market-financed euro area-wide investment plan to be agreed, monitored and even administered by representatives of northern governments, should be acceptable to them and would support the southern ones in their endeavor of risk reduction.

Keywords Globalization · Populism · Risk reduction · Risk sharing · Fiscal austerity · Investment plan

1 Introduction

The euro area crisis was ignited by the Greek shock and seemed to prove that the northern fear of a southern fiscal profligacy, dating back to the beginning of the process of monetary integration, was fully justified (although such a conclusion ignored the fact that until the beginning of the crisis, public finances of the countries mostly affected were in relative good shape or at least were giving signs of convergence). Still, fiscal austerity was imposed in order to redress the situation, although it was only thanks to the reassurance coming from the ECB, acting de facto as a lender of last resort, that the crisis came to an end.

P. Della Posta (✉)
Università di Pisa, Pisa, Italy
e-mail: pompeo.della.posta@unipi.it

L. Paganetto (ed.), *Capitalism, Global Change and Sustainable Development*, Springer Proceedings in Business and Economics,
https://doi.org/10.1007/978-3-030-46143-0_2

A different view results from a southern perspective, arguing that northern countries are benefitting from the rigidity provided by EMU, since their trade surpluses do not find any automatic compensatory mechanism to balance them.

Both perspectives seem to lead to the conclusion that creating the EMU was not appropriate and that it would be in the convenience of southern countries to leave and in the convenience of northern countries to let them go.

This conclusion, however, ignores the deeper reasons that are behind the creation of EMU, which is part of the wider project of European integration.

If northern countries want the southern ones to redress their fiscal situation, thereby reducing risk, as the current fiscal rules would oblige them to do, then, considering their self-interest in not disrupting the whole project of European integration, they should be willing to accept some degree of risk sharing in order to counterbalance the negative effects of risk reduction, for example in the form of the market financed euro area wide investment plan proposed by Della Posta et al. (2018).

2 Euro Area Crisis, Austerity Policies and the Stabilizing Role of the ECB

The euro area crisis is usually attributed to public debt imbalances (this is also why it is often defined as "sovereign debt crisis", to distinguish it from the "global financial crisis" that preceded it). This is a debatable conclusion, though, especially considering the role also played by private debt, by current account imbalances and by self-fulfilling expectations, ignited not only by the sudden discovery of the higher than expected Greek public deficit and debt, at the end of 2009, but also by the "Deauville declaration" with which in October 2010 Angela Merkel and Francois Sarkozy acted as "scaremongers of first instance" (the opposite of what a "lender of last resort" should be doing in order to reassure the markets), thereby fostering rather than extinguishing the crisis.

Moreover, before the crisis the public debt-to-GDP ratio had been gradually decreasing in all euro area countries, and it was rather low in Spain and Ireland (respectively about 40% and 20%). It is only after the Greek crisis and after the bursting of the housing bubble in some countries (Spain and Ireland, where the bubble was induced by an excess of private debt made possible also by the capital inflows from northern countries, particularly France and Germany), that public debt had to be increased in order to bail out the private one.

In the case of Portugal, instead, the problem had not to do with a house bubble, but with current account imbalances (also present in the case of Spain, Ireland and Greece), once more resulting from an excess of consumption made possible on one hand by the capital inflows attracted by the higher interest rates and the perception of a rather low risk in southern countries, and on the other hand by the rigidity characterizing monetary unions and preventing any automatic adjustment of the current account imbalances (this point will be further addressed below).

Be that as it may, being the crisis attributed, by Germany and other northern European countries, to public debt imbalances, namely being it believed to be fundamentals-driven (the main economic fundamental to be monitored, however, being public debt and not the private one), the proposals for its resolution have been based mainly on the imposition of fiscal austerity policies, in order to adjust such imbalances.

The counterproductive effects of those policies had been immediately recognized by many observers (including Krugman 2010), who had applied the basic Keynesian model to conclude that with their adoption the public debt to GDP ratio would have increased rather than decreased because of the contractionary effect they would have had on GDP, the denominator of the public debt-to-GDP ratio.

As a matter of fact, as De Grauwe and Ji (2013) had promptly shown, the public debt-to-GDP ratio did not fall, and in fact it kept increasing, in spite of fiscal austerity

It is undeniable, then, that rather than thanks to fiscal austerity, the crisis came to an end only when the ECB stepped in and reassured the markets by acting, eventually, as a "lender of last resort". This happened with the famous Draghi's "whatever it takes" speech (accompanied by the prompt adoption of the unlimited Outright Market Transactions program), that reduced dramatically the interest rate on public debt.

Such an ECB role, however, like any other form of implicit or explicit fiscal solidarity, had been denied that far, following too literally the prescriptions of the Maastricht Treaty, with the motivation that doing otherwise would have risked providing the wrong incentives to the market. This is why the little Greek fire was allowed to grow (an image that I am borrowing from Paul De Grauwe), putting at risk the whole of Europe.

The role of incentives is well-known in economics and their importance cannot be discarded. It should be also recognized, however, that there are instances in which they may play a rather limited role, for example if divergences are structural rather than dependent on the deliberate actions of agents. This is the case, for example, when countries differ in labor or market institutions, or when the public debt that they have inherited is larger than that of other countries, thereby implying the need to impose higher taxes (especially when considering the indirect ones, like VAT) to repay it. This would undermine the domestic competitiveness since, in such a case, countries belonging to a monetary union would be characterized by different inflation rates, in spite of the same monetary policy being run in the area by the common central bank. Another instance is when countries have a different productive structure and the process of economic specialization based on comparative advantages has implied concentrating in a sector whose productivity increases less than that of other sectors. A further problem exists if, as argued above, no self-stabilizing market mechanisms are allowed to be in place by the system that has been adopted.

3 Opposing Views on the Way EMU Should Be Run: Exit EMU, Risk Reduction or Risk Sharing in EMU?

3.1 *Exit EMU from a Southern Perspective*

A quite wide spectrum of economists of southern euro area countries, both of a progressive and a conservative orientation (in Italy, for example, including names like Alberto Bagnai, Domenico Mario Nuti and Paolo Savona, just to list some of the most well-known of them) observe that the adoption of the euro has coincided with the relative impoverishment of their countries and with the enrichment of the Northern ones, therefore arguing that their countries should leave EMU, which is for them a sort of straightjacket that does not allow them to operate freely and that forces them in a subordinate role vis-à-vis the northern ones.

This is a long standing argument, that was raised already when the European Monetary System was created. After the fall of the fixed exchange rates system created in 1944 in Bretton Woods, European currencies started fluctuating widely against each other, thereby disrupting the otherwise intense intra-European trade. The late Marcello de Cecco also observed, however, that an additional effect of floating exchange rates within Europe would have been the appreciation of the German mark vis-à-vis the other European currencies. As a matter of fact, the economic strength of Germany would have attracted capital from the rest of the world, thereby appreciating the D-Mark and reducing its commercial competitiveness. Moreover, any German competitive advantage over its trading partners, resulting in a trade surplus, would have created the automatic conditions for rebalancing the situation through the appreciation of its currency, as it was the case when the "rules of the game" in the Gold Standard were not infringed by Britain's sterilization policies. Joining a fixed exchange rate system, then, this was the argument made by Marcello de Cecco, was in the full interest of Germany, since it allowed that country to avoid the appreciation of the D-Mark that would have prevented the accumulation of current account surpluses. The same is true, of course, for the adhesion to a monetary union, which is an extreme form or irrevocably fixed exchange rates.

German economists and citizens, then, should consider the role played by the mechanisms described above if they want to understand the true origins of the current account deficits of most southern euro area countries and, as a result, the origins of their persistent current account surpluses.[1]

---

[1] When discussing the role played by the European Monetary System, however, it also needs to be recalled that, according to an old argument introduced by Giavazzi and Pagano (1988), the German monetary discipline might have helped reinforcing the anti-inflationary credibility of follower countries, including Italy, so as to contribute to the reduction of their inflation rate. Such a conclusion, however, is subject to several caveats and discussions that cannot be reported here but that should be considered.

3.2 Exit EMU from a Northern Perspective

The appeal signed by 154 German economists and published in the Frankfurter Allgemeine Zeitung on May 21, 2018 against Macron's proposal to transform the European Stability Mechanism to a permanent European Monetary Fund in order to be able to absorb future economic shocks, goes in the opposite direction.[2] The most prominent of those signatories are Hans-Werner Sinn, former president of the Ifo Institute, and Jürgen Stark, former member of the European Central Bank executive board. Their views are also in line with the position expressed by another well-known German economist like Bernd Lucke, who has been among the first ones to interpret and represent the German sentiment of fear of the mutualization of southern public debt. A similar fear characterizes the appeal signed in March 2018 by eight northern euro area countries (Finland, Denmark Sweden, Lithuania, Latvia, Estonia, Ireland and The Netherlands).

The northern view resulting from those documents only focuses on the risk they face to be asked to pay the debt of southern countries. This is why they encourage the states that are not capable to keep their accounts in order to leave EMU and to make this possible they suggest to set up a procedure to be followed by insolvent countries to quit the euro.

This position might convey the idea that they do not fear the risk that some countries may be leaving the euro area. What is not clear, however, is whether they would be willing to accept the risk that such an exit process involved more than just an outsider country like Greece, and whether they are well aware of the effects and implications this would imply in terms of the overall process of European integration.

One possibility is also that they rely on the fact that a deflagration of the European Union might not be in the interest of follower countries either, since this would mean to have to cope with a newly isolated and strong Germany, precisely what European countries wanted to avoid when EMU was created.

3.3 Risk Reduction and Risk Sharing (Namely Fiscal Solidarity) in EMU and the Benefit for Northern Countries of Being Part of It

Opposite, although less radical positions are those that, at least in principle, do not question the participation in EMU, although they subject it to the fulfillment of either risk reduction (to be performed by southern countries) or risk sharing (to be provided by the northern ones).

[2]https://www.faz.net/aktuell/wirtschaft/konjunktur/oekonomen-aufruf-euro-darf-nicht-in-haftungsunion-fuehren-15600325.html?utm_term=0_10959edeb5-f3892c4dc9-189813013&utm_campaign=f3892c4dc9-

Risk reduction is required by northern countries, arguing that if South and North have to remain together in a monetary union, the former has to behave properly, not expanding public finances in the implicit hope of a northern bail out. Such risk of fiscal profligacy is something that has been present in the debate from the very beginning of the process of monetary unification, namely from the Delors Report (Delors 1989).

It is far from clear, though, why risk reduction should only apply to budgetary issues. This is what has emerged after 2010, with the adoption of the European Semester, according to which a whole set of potential macroeconomic imbalances has to be monitored by the European Commission. As a matter of fact, not only fiscal divergence might represent a potential risk for EMU, but also any other indicator of competitiveness divergence could. Introducing labor market reforms in an uncoordinated manner (a clear example being the German Hartz reforms) or adopting a preferential fiscal treatment in favor of foreign companies, also creates a divergence within EMU, which increases the risk of a long run unsustainability. The latter case, for example, in which countries act the facto as free riders (this is the case of Ireland and the Netherlands, for example), produces the paradoxical result of strengthening the public finances of the countries adopting such fiscal measures, while weakening the public finances of the countries not adopting them, whose revenues will be decreasing and creating an equally disruptive trade divergence within EMU because of the competitive advantage it gives them.

Risk sharing is required instead by southern countries, arguing that if South and North have to remain together in a monetary union, the latter should acknowledge and give back to the former at least part of the advantage gained with the creation of the monetary union.

As a matter of fact, the euro area crisis itself would have been avoided by the application of fiscal solidarity, that has been exerted only at a later stage, and only being subject to conditionality in order for northern European countries to be reassured of not being exploited by the southern ones.

If the two perspectives reported above—the northern and the southern one—are brought to an extreme, the conclusion would be the one presented in the previous section, namely that southern countries that are not capable to provide the required risk reduction or that are going to lose from their participation in EMU because of the permanent competitive advantage that export countries have, should leave EMU.

4 The Need for a Comprehensive View: Risk Reduction and Risk Sharing

In my view both extreme positions make the mistake of looking at only one aspect, assigning to it an exorbitant weight that overcomes all other aspects and reasons for the creation of the euro, that should be considered instead in order to have an overall, balanced view on being part of EMU.

Those reasons might have to do, for example, as recalled above, with the high inflation rate of the 1970s and part of the 1980s, that was reduced also with the help (although not without costs, as the rational expectations theories of those days would have suggested instead) of the fixed exchange regime that had been adopted, and that set the adhering countries on a sustainable path of credibility. Needless to say, it is easy today to forget about inflation, given its current low level.

Similarly, we have forgotten the consequences of exchange rate volatility for an economic area open to trade and characterized by many different currencies and in which the suspect of "beggar thy neighbor" policies induce competitive devaluations and produces a spiral of retaliatory measures. Would that problem disappear if we leave the euro? May be that this is the case, since after all world trade operates under a flexible exchange rate regime. The intensity of such trade, though, is much higher in Europe. Not to mention that we came to the conclusion that we need intra-European trade and open commercial relationships, in order to keep away war and the phantoms of the past.

Maybe even more importantly, we have forgotten that the euro was part of a larger design, aiming at the political unification of the European continent. True, that might have been risky (as the Italian politician Carlo Ripa di Meana observed in discussing the project of EMU: "It is not the European Union that creates money but money that should be creating the European Union: it's a risky game"), but the final objective of a European Union, after the dramas of two world wars, as Altiero Spinelli and the other authors of the Ventotene Manifesto had dreamed, was the main ideal motivation behind it and the previous moves, the creation of the customs union in 1957 with the Treaties of Rome and the adoption of the Single European Act in 1987 had shown that the neo-functionalist approach had been giving good results.

Finally, behind the beginning of the process of European integration, and indirectly behind EMU, then, there is the need for European countries to cooperate to better resist the pressures of globalization and protect their citizens, being aware that this would only be possible by joining forces and by sticking together.

Still, in order to reap a (limited) advantage (or to avoid incurring a limited loss), both northern and southern countries criticize EMU by focusing on a single aspect and by neglecting the risk to ignite the breakup of the process of European economic, monetary and political integration, which would be a much worse and potentially disruptive conclusion.

In other words, this implies the risk to play a prisoner's dilemma game, in which both northern and southern European countries do not resist the temptation to play alone in the hope to reap a small benefit, not realizing that by doing so they will end up in a Pareto inferior equilibrium.

5 Negative Implications of Austerity Policies and the Need for an Investment Plan

The austerity policies that have been applied in order to solve the "sovereign debt" crisis have implied a dramatic fall of investments—the only expenditures that can be cut easily by policymakers interested to preserve an electoral consensus without receiving any objections or complaints on the part of those who would have benefitted from them, namely the voiceless future generations.

Most crises of the past, however, have been put to an end by following an opposite route: economic growth has been pursued in order to reduce the public debt-to-GDP ratio, thanks to the beneficial effects of the Keynesian multiplier, so that even if the numerator would be growing further, the denominator would grow even more. This has been the case, after World War II, in Britain with its 240% public debt-to-GDP ratio, or in the USA, where GDP growth has been the key element to stabilize and make public debt sustainable.

The least that this may imply is that the austerity policies followed by southern euro area countries, should have been supported and partially compensated in a cooperative environment in order to avoid the GDP fall of fragile countries that would nullify their effort, by expansionary domestic policies to be undertaken by the northern euro area members. Such a compensatory (not necessarily direct, through fiscal transfers from North to South countries, but also indirect, through higher internal absorption—consumption, investment, government expenditure—that would have favored a rebalancing of North-South current account divergences) would be fully beneficial because it would increase the GDP growth of the most fragile countries, so that the public debt/GDP ratio will surely decrease. Since this was not done, no surprise that the Greek paradox emerged, implying that not only austerity policy imposed a heavy social cost (that became apparent also in the mortality rates, as documented in the literature), but the public debt to GDP ratio did not fall, due to their contractionary effect on GDP growth.

It could still be understood that this would be hard to accept by citizens who speak another language, have a different religion, different traditions, culture and history, and different beliefs, although this signals clearly that they do not feel being all Europeans.

What is harder to understand, though, is the fact that solidarity if not accepted even if its cost would be close to zero. As a matter of fact, Germany and other northern euro area countries would only be required, following for example the proposal of the adoption of a euro area wide investment plan made by Della Posta, Marelli and Signorelli, to join forces with the southern euro area ones and issuing European bonds to collect the necessary financing on the international markets. As a matter of fact, the guarantee by euro area countries would be sufficient to attract market capital, so as to enjoy a high leverage effect. Moreover, still according to the proposal made by Della Posta et al. (2019), for example, investments would be selected, administered and monitored by the countries who provide a guarantee on them. Such a solidarity would imply neither a financial nor a monetary sacrifice, given the eagerness with which international financial markets would be looking for additional ways to diversify their portfolios.

Similarly, expansionary and growth enhancing proposals, contributing therefore to the reduction of the public debt-to-GDP ratio, are reported in Discussion Group: "Growth, Investment and Territory" (2018).

The Juncker Plan (implying investing up to 2% of GDP in four years in the economies under stress) has done something similar, having been administered through the EIB (European Investment Bank). Such a plan should be greatly relaunched and reinforced.

6 Fighting Against Populism

It deserves to be underlined that the proposed policy would have a further very relevant implication, namely fighting against populism. As a matter of fact, the fiscal austerity policies adopted in response to the euro area crisis have negatively affected the sentiment towards Europe in southern countries. Adopting an expansionary, market-financed investment plan, then, promises to produce a positive effect not only on GDP growth, on interest rates and on public debt-to-GDP ratios, but also—and maybe even more importantly—on the perspectives themselves of the long run success of EMU and potentially of the EU.

It could be argued that while this would gain the favor and the consensus of southern euro area countries, the opposite would be true for the citizens of northern Europe. But they should be reminded that the true choice they have to make is not between accepting or not accepting a big Pan-European, market financed and growth enhancing investment plan, but rather between accepting the investment plan or having to face a situation in which the euro area breaks up, since no country will desire to remain in a union in which the costs exceed the benefits and not protecting its citizens, contrary to what initially promised.

In other words, they risk playing with southern countries the well-known prisoner's dilemma game, leading to an undesired to both but inevitable Pareto inferior Nash equilibrium.

7 Concluding Remarks

The euro area crisis was ignited by the Greek shock and seemed to prove that the northern fears of a southern fiscal profligacy, dating back to the beginning of the process of monetary integration, were fully justified, in spite of the fact that until the beginning of the crisis the public finances of the countries mostly affected were in relative good shape or at least were giving signs of convergence.

In spite of that, fiscal austerity was imposed in order to redress the situation, but it was only thanks to the reassurance coming from the ECB, acting de facto as a lender of last resort, that the crisis came to an end.

From a southern perspective it is often argued that northern countries are benefitting from the rigidity provided by EMU, since their trade surpluses do not find any automatic compensatory mechanism to balance them.

Both perspectives seem to lead to the conclusion that creating the EMU was not appropriate and that it would be in the convenience of southern countries to leave and in the convenience of northern countries to let them go.

This conclusion, however, ignores the deeper reasons that are behind the creation of EMU, which is part of the wider project of European integration.

If northern countries want the southern ones to redress their fiscal situation, then, as the current fiscal rules would oblige them to do, considering their interest in not disrupting the whole project of European integration they should be willing to accept some risk sharing. In this paper, for example, I refer to the proposal of a euro area wide investment plan agreed, monitored and even administered by representatives of the governments of northern euro area countries and financed by Eurobonds (namely by money collected on the market and not through the taxes of euro area citizens. In such a way, risk sharing and solidarity would be exerted without actually asking northern countries any financial disbursement, something that should be acceptable for them and that would help the southern ones in their endeavor of risk reduction.

References

De Grauwe, P., & Ji, Y. (2013). From panic-driven austerity to symmetric macroeconomic policies in the Eurozone. *Journal of Common Market Studies, 51*(Annual Review), 31–41.

Della Posta, P., Marelli, E., & Signorelli, M. (2018). A market-financed and growth-enhancing investment plan for the euro area. Presented at the 59th Annual Scientific Meeting of the Società Italiana degli Economisti, Bologna. *Globalizzazione e sviluppo: città, regioni, nazioni*, October, 25–27, 2018.

Della Posta, P., Marelli, E., & Signorelli, M. (2019). An immediate solution for the euro area crisis: A grand european investment plan. In L. Paganetto (Ed.), *Yearning for inclusive growth and development* (pp. 113–135). Good Jobs and Sustainability: Springer.

Delors, J. (1989). Report on economic and monetary union in the European community. Presented April 17, 1989 (commonly called the Delors Plan or Report) By Committee for the Study of Economic and Monetary Union.

Discussion Group: "Growth, Investment and Territory". (2018). *The response to the needs of the citizens and the turnaround in European policies: a common cultural and political platform for a radical change in European economic policies*, December, by Riccardo Cappellin, Enrico Ciciotti, Maurizio Baravelli, Raffaele Barberio, Leonardo Becchetti, Marco Bellandi, Fiorello Cortiana, Fiorenzo Ferlaino, Franz Foti, Gioacchino Garofoli, Enrico Marelli, Carlo Pescetti, Luciano Pilotti. http://economia.uniroma2.it/dmd/crescita-investimenti-e-territorio/

Giavazzi, F., & Pagano, M. (1988). The advantage of tying one's hands: EMS Discipline and Central Bank Credibility. *European Economic Review, 32*(5), 055–088.

Krugman, P. (2010, July 7). Self-defeating austerity. *The New York Times*, The Opinion Pages, The Conscience of a Liberal. Retrieved from http://krugman.blogs.nytimes.com/2010/07/07/self-defeating-austerity/?_r=0

Global Integration and Economic Growth in Emerging Countries: The Case of BRICS and NEXT-11

Misbah T. Choudhry, Enrico Marelli, and Marcello Signorelli

Abstract This paper investigates the impact of global integration on economic growth in a large group—BRICS and NEXT-11—of emerging countries and it tries to verify a possible different relation for the second group. The period of analysis is 1980–2015. Our hypothesis is that the impact of global integration (measured as foreign direct investment and share of trade as percentage of GDP) on economic growth is not only direct but also indirect through various other determinants of economic growth. Thus, by using panel data econometric estimation techniques, multiplicative models are estimated. Results show that global integration—both trade openness and FDI inflow—benefits economic growth. The coefficients are however higher in the BRICS group rather than in the complete sample.

Keywords Globalisation · Global integration · Economic growth · Trade openness · FDI · Emerging countries · BRICS · NEXT-11

JEL P52 · P33 · F14 · O53

1 Introduction

Globalization favoured, particularly in the last half century, the development of many previously "underdeveloped" or developing countries. The fast growth of the BRICS group, in particular China and India, is just the clearest example (while

M. T. Choudhry · E. Marelli · M. Signorelli (✉)
Centre for Research on Economic Empowerment of South Asian Women, Luton, UK

Department of Economics and Management, University of Brescia, Brescia, Italy

Department of Economics, University of Perugia, Perugia, Italy
e-mail: misbah@creesaw.org; enrico.marelli@unibs.it; marcello.signorelli@unipg.it

L. Paganetto (ed.), *Capitalism, Global Change and Sustainable Development*, Springer Proceedings in Business and Economics,
https://doi.org/10.1007/978-3-030-46143-0_3

growth has been more volatile in Brazil, Russia and South Africa).[1] NEXT-11 is another group of fast growing countries.[2] Many other groupings of emerging countries can be found in the literature and in the reports by the World Bank and other international organizations.

Most economists generally view globalization in a positive way. As a matter of fact, income inequalities have been reduced in the world, although they have certainly increased within countries, at least in the developed world. A too rapid globalization may induce the excessive exploitation of both natural and human resource (also with the risk of dreadful "social dumping" events). We also know that fast economic liberalisations, especially in the financial sector, with lacking or weak international regulations, have been—and still may be—the cause of financial crises. So globalization, especially if unregulated, is not always the best policy option (see Stiglitz 2002; Acemoglu et al. 2003; Rodrik 2006).

Nevertheless, globalization has been mostly beneficial to developing countries. This paper investigates, the impact of global integration on economic growth in a large group—BRICS + NEXT-11—of emerging countries and it tries to verify a possible different relation for the second group. The period of analysis is 1980–2015. Our hypothesis is that the impact of global integration (measured as foreign direct investment and share of trade as percentage of GDP) on economic growth is not only direct but also indirect through various other determinants of economic growth. Thus, by using panel data econometric estimation techniques, multiplicative models are estimated—which is an original feature of this paper—to capture the contemporaneous impact of global integration as well as other determinants of GDP per capita income growth.

We remark here that empirical investigations regarding the NEXT-11 group (as a whole) are extremely rare. Our results show that global integration—both trade openness and FDI inflow—benefits economic growth. The coefficients are however higher in the BRICS sample rather than in the complete sample (BRICS + NEXT-11), probably because in the countries of the NEXT-11 group with a lower development level the benefits of trade liberalisations cannot be fully exploited.

The structure of this paper is as follows. After this Introduction, we review both the literature on the links between trade opening and economic growth, then the growth effects of FDIs. In Sect. 3, we present our econometric model, with the specification of the estimation methods. Data are presented in Sect. 4, that includes also some descriptive statistics. Econometric results are presented and discussed in Sect. 5. Section 6 concludes.

---

[1]Notice that exports of Russia and Brazil are much more dependent on raw materials and energy sources. The BRIC acronym was first employed by O'Neill (2001).

[2]It comprises Bangladesh, Egypt, Indonesia, Iran, Mexico, Nigeria, Pakistan, the Philippines, Turkey, South Korea, and Vietnam. These countries could potentially become some of the worlds' largest economies.

2 Literature Review

Economic growth depends on several factors. Classical and neoclassical economic theories considered capital accumulation as a key driver of growth, while more recent growth models highlight the role of human capital.

Capital stock formation depends on both private and public fixed investment decisions, which can be supported by a high saving rate, but partly derives also from foreign direct investment. So, FDI and fixed capital stock formation can enter into growth equations as independent variables.

On the other hand, extended endogenous economic growth models (Grossman and Helpman 1990, was the first of a long series) provide a role for trade opening: growth depends on the rate of knowledge accumulation and technological progress, which is stimulated by (domestic and international) competition, hence by trade liberalisation policies. In particular, developing countries not only can access to intermediate and high-tech goods (so important for economic growth) through imports, but they can also acquire valuable foreign reserves through exports, thus making sustainable the mentioned imports. These theoretical underpinnings have favoured the adoption of trade liberalisation policies in many developing countries (Krueger 1978).

However, we must add that trade openness can be potentially detrimental to economic growth for developing countries; for example, when the latter tend to specialise in traditional industries or in sectors where R&D activities are not the core ones (see Rodríguez and Rodrik 1999, 2001). Furthermore, countries may also become vulnerable to negative external shocks, boom-bust cycles of investment, volatile exchange rates, dumping, imported inflation (see Montalbano 2011; Iyke 2017).[3] Some studies found an income threshold below which more openness deters the growth prospects, the key reason concerning the human capital constraints that limit the benefits of international technology transfers (Kim and Lin 2009).

Notice that most empirical studies focused on the links between trade openness and economic growth, while more recent investigations regard the relations between FDI and growth. However, trade and FDI are also interrelated, since FDIs are often "export-oriented", that is they establish productions partly (in some instances mostly) oriented to exports.

Although we shall present the literature review separately for the trade and the FDIs determinants of growth—consistently with the empirical studies focusing on one variable at a time—there are some researches considering them jointly. This is also the case of our paper.[4]

---

[3]A negative impact of trading opening may be caused by the worsened income distribution. Nayyar (2015) argues that policies that have stressed more openness in trade, investment and finance, have dampened output growth through a deteriorated income distribution.

[4]In addition to some studies reviewed in Sect. 2.1; for example, Cieślik and Tarsalewska (2011), Marelli and Signorelli (2011), Sulochani and Leeb (2015).

2.1 *The Links Between Trade Openness and Economic Growth*

The empirical works on the growth effects of trade have followed a variety of approaches (Dollar 1992, was one of the initial studies). As far as the dependent variable is concerned, following the main literature (e.g. Frankel and Romer 1999), almost all empirical studies chose the GDP per capita expressed in growth rates. More discussion arose relative to the key explanatory variable: trade openness.

A first important distinction refers to the choice between analysing trade growth—whatever is the driver (liberalisations but also reduction in transport and communication costs or other reasons) and explicit trade policy: liberalizations, reduction of duties or non-tariff barriers, participation to trade agreements, etc. Some studies (e.g. Edwards 1993; Milner et al. 2007), in order to assess the effects of liberalizations and institutional change, compute sophisticated indices of trade openness, based on tariffs, non tariff-barriers, degree of effective protection, etc.

Regarding this approach, we recall the sceptic's view of Rodrìguez and Rodrik (1999), who find "little evidence that open trade policies—in the sense of lower tariff and non-tariff barriers to trade—are significantly associated with economic growth"; the reason is that "the nature of the relationship between trade policy and economic growth remains an open question".

A second relevant issue refers to the measurement of trade openness (see also Sakyi et al. 2015, for a review of the alternative measures). The most frequently used variable is the total trade share in GDP, i.e. (exports + imports)/GDP. Sometimes the export share or import share alone are considered. Furthermore, instead of the "nominal trade shares", in some studies the "real trade shares" are computed, where the denominator is GDP adjusted for PPP. The nominal trade share are also accused to be biased, since they overestimate the openness degree of small trading countries (and vice versa). Thus, Squalli and Wilson (2011) propose "composite trade shares", that take into account to what extent an open economy trades internationally and how much it is a substantial contributor to world trade; in other words, it considers not only a country's share of trade (on domestic GDP), but also its interaction and interconnectedness with the rest of the world.

A more decisive critique to traditional measures of openness is raised by Sulochani Ramanayakea and Leeb (2015). They show—in a study comprising 156 developing and 49 developed countries for the period 1980–2009—that export growth is the most robust measure, in addition to export specialization, compared to traditional variables of trade openness and FDI; in fact, export growth requires capability building in indigenous firms and investments in innovations. Moreover, productivity increases are guaranteed through "learning-by-exporting" mechanisms only if export growth is maintained over time. This assumption is consistent with the evidence that some developing countries show growth surges for a certain period of time, but are unable to sustain this growth over a longer period (see also Rodrik 2006).

From an econometric point of view, recent studies[5] follow either cross-section or panel approaches, usually with large samples of countries: see for example Edwards (1998), Makki and Somwaru (2004), Sarkar (2008). The distinction between developed countries and emerging economies has become more common in recent works, also because globalization has led most emerging countries to open their economies. Some specific studies are devoted to transition countries[6]; Iyke (2017), by using fixed-effects regressions for 17 CEE (Central and Eastern European) countries over the period 1994–2014 and an index of trade openness similar to the one proposed by Squalli and Wilson (above illustrated), find that trade openness is important for real GDP per capita growth. A similar positive link between trade liberalisation and growth for this group of countries is also detected by Nannicini and Billmeier (2011).

In this paper, we focus particularly on emerging economies. As already anticipated, it is in these regions of the world that some countries are unable to exploit the benefits of technology spillovers and so trade openness is not beneficial to growth. Hence, the link between trade openness and economic growth is not always significant or robust. In some cases, it can even be negative: Kim (2011), by using the instrumental variable threshold regressions approach, finds that greater trade openness has strongly beneficial effects on growth and real income for the developed countries, but significantly negative effects for the developing countries.

The structure of trade is also important to determine the final result; in fact, it has been found that developing countries specializing in manufacturing are the ones benefiting more from trade and achieving higher economic growth. Even the development of the financial sector can strengthen the link between trade openness and economic growth; Huang and Chang (2014), using panel data for 46 countries from 1983 to 2007, find that it is especially in countries with higher stock market development that more trade openness enhances economic growth. However, Herwartz and Walle (2014), using data from 78 economies for the period 1981–2006, interestingly find that very high levels of financial openness generally erode the growth-promoting role of financial development while high trade openness strengthens it.

Recent empirical research on this topic normally refers to a large number of developing countries. For instance, Cieślik and Tarsalewska (2011), analyzing a group of 97 developing countries in the period of 1974–2006 and using static and dynamic panel data estimation methods, find that both international trade and FDI positively contribute to growth; this study is also worth to be mentioned because, differently from most works, derives the equations to be estimated from a theoretical model.

[5]For previous works, see the extensive review of Edwards (1993). A recent survey is the one by Singh (2010).

[6]These countries have some specific features; for instance, their liberalisations and opening to trade have been much faster (the so-called "shock therapy" adopted in the '90s) compared to the more gradual approaches of China and India (see Marelli and Signorelli 2010).

Some studies focus on a very limited number of countries.[7] Marelli and Signorelli (2011) investigate the case of China and India: the effect on economic growth (in terms of GDP per capita) of both trade openness and FDI is found to be positive, statistically significant and robust; here, the reverse causality issue is tackled with a 2SLS estimation. Mercan et al. (2013) confirm the positive and significant effect of openness on economic growth for a group of five countries (BRIC plus Turkey) in the period 1989–2010. In the literature, it is possible to find several studies focusing on individual countries, such as India, Mexico, Pakistan, South Korea, and many others.

Almost all econometric investigations include in the estimated equations some control variables. In some studies, the initial level of income is included among the explanatory variables, in order to analysis the "converge" of countries. The most common control variables include, however, total investment (gross capital formation), life expectancy, size or growth rate of population, human capital, R&D and innovations, size of firms, patent protection, credit, inflation rate, index of economic freedom, political institutions, government expenditure, export specialisation, and some others.

The role of human capital can be particularly relevant; in fact, the benefits from trade openness are adequately exploited if a country has adequate absorbing capacity of new technology, that depends on the level of human capital. Jadoon et al. (2015), by considering the case of eight Asian economies,[8] further distinguishing between higher income and lower income countries, find that the impact of trade openness on human capital has been significant only for the formers (although both groups of countries enjoyed the trade-led growth).

A typical econometric problem encountered in econometric estimations is the issue of endogeneity, or reverse causality. Frankel and Romer (1999) were among the first to use instrumental variables to overcome it. In most instances, a GMM estimation is the solution. Difference-in-difference estimations are also common. Panel unit root tests are frequently used. The cross-country heterogeneity problem is generally faced by considering a sample of relatively homogenous countries.

The time-series approach is uncommon (see also Harrison 1996). Xu and Wang (2007) use time-series estimating a system of four equations. Ramjerdi (2007) explain Chinese growth in terms of capital and technical progress; the effects of opening up are examined indirectly by splitting the period into two sub-periods (before and after 1978, when the "open door" policy started).

Finally, some other studies follow cointegration or VAR approaches. We just mention[9] Sakyi et al. (2015), who investigate a large sample of 115 developing

[7]In several cases, cross-sections are based on sub-national data, e.g. China's provinces: see Sun and Parikh (2001), or on many sectoral data: see Milner et al. (2007).

[8]Lower income countries considered in the study are: India, Indonesia, Pakistan and Sri Lanka; higher income countries: Japan, Malaysia, Singapore and South Korea. The period is from 1981 to 2012.

[9]Other works include Liu et al. (2002), Sahoo and Mathiyazhagan (2008), Zhao and Du (2007); while Tsen (2006) focused on Granger causality.

countries for the period 1970–2009, further distinguishing three mutually exclusive groups of countries: low-income, lower middle-income, and upper middle income countries. By employing non-stationary heterogeneous panel cointegration techniques, they find a positive bidirectional relationship between trade openness and income level, although the link between openness and income is higher for upper middle-income countries.

The general conclusion of most empirical studies is that the link between trade openness and economic growth is positive, significant and robust. The exceptions refer mainly to less developed countries (see Kim and Lin 2009, as a good example of the negative link).

2.2 *Growth Effects of FDIs*

The interdependence between FDI inflow into host countries and the economic growth has been subject to various research projects for years. The general belief suggests that FDI has contributed to economic growth in developing countries through direct and indirect channels. Notice that FDI is not only a major source of technology and know-how for developing countries,[10] but also a key foreign financing font. The positive impact of FDI inflow in a host country reveals itself, more generally, as capital accumulation, technology transfer, know-how acquisition, innovative capacity and ultimately, economic growth. World Investment Report (2002) suggested that several studies prove that FDI can promote economic development of the host country, also by favouring exports of the country. However, it should be added that the relationship between foreign Multinational Corporations and their host countries varies and the effect of FDI on economic growth is chiefly dependent on the policy environment.

Therefore, most developing countries have been trying to increase the FDI inflow within their country due to the mentioned general belief, although the findings of empirical research on this topic vary. Some studies, such as Choe (2003), Yao (2006), Mullen and William (2005) share opinions with Anwar and Nguyen (2010), who state that FDI not only increases the supply of capital but it can also buttress technology transfer. On the other hand, some authors, such as Carkovic and Ross (2005) or Adams (2009), conclude that FDI has no significant impact on economic growth. Others, such as Borensztein et al. (1998), find little support for the correlation. Alfaro et al. (2004) conclude that FDI alone plays an uncertain role and that FDI entry could lead to positive economic growth when only fundamental factors are in play.

The consideration of additional control variables can make the difference also in this case (additionally to what illustrated for the links between trade openness and

[10]In particular, general wisdom is that multinational firms sustain, through FDI, the international knowledge diffusion.

economic growth). For example, Batten and Vo (2009) show that FDI has a stronger positive impact on economic growth in countries with a higher level of education attainment, openness to international trade and stock market development, and a lower rate of population growth and lower level of risk.

Now we shall briefly illustrate the results of empirical investigations regarding some individual developing countries. We begin with the studies where the link between FDI and growth results to be positive and significant, then we analyse research where the relation turns out to be absent, weak or even negative.

A first interesting case is that of China. Tang et al. (2008) analyze the data from 1988 to 2003 and, using an integration and Granger casuality analysis, found causality running from FDI to GDP in China.

A study based on simultaneous equation model, using panel dataset for 61 provinces of Vietnam, reveals that in overall terms a mutually reinforcing two-way linkage between FDI and economic development exists in Vietnam (Anwar and Nguyen 2010). The study also suggests that the impact of FDI on development is related to the amount of resources that are invested in education and training, financial market development and the capacity of reducing the technology gap between the foreign and local firms.

In another study (Narayanamoorthy et al. 2009) the casual relationship between FDI and growth of BRICS countries was examined; the results show that economic growth leads FDI bi-directionally for Brazil, Russia and South Africa and FDI leads growth uni-directionally for India and China respectively.

In some studies, the sign and significance of the link between the two variables depend on the economic sectors and other economic conditions. Chakraborty and Nunnenkamp (2008) assess the relationship between FDI and economic growth in India, by using industry specific FDI and output data; they apply Granger causality tests within a panel cointegration framework. The results prove that growth effects of FDI vary widely across sectors: the casual relationship is absent in the primary sector whereas in the manufacturing sector the relationship is strong. Though the effects were only transitory in the services sector, FDI in the services promotes growth in the manufacturing sector through cross-sector spillovers. The economic reforms of 1991 attracted booming FDI into the country to foster growth. Yet, earlier studies on India have failed to find a significant growth impact (e.g., Agrawal 2005; Pradhan 2002).

Similarly, Khaliq and Noy (2007) find that FDI's impact varies over the different sectors in Indonesia. The authors conclude that, at an aggregate level, FDI is observed to have a positive effect on economic growth; whereas, when accounting for the different average growth performance across sectors, the beneficial impact of FDI is no longer apparent. In fact, few sectors showed a positive impact of FDI on growth and one sector, mining and quarrying, even shows a negative impact.

In another study, Yalta (2013) explore the casual relationship between foreign direct investment and GDP in China for 1982–2008, both in bivariate and a multivariate framework. The results show that a statistically significant relationship between FDI and GDP does not exist. The author argues that the effects of capital accumulation and positive knowledge spillovers, predicted by endogenous growth models, may not occur in developing countries due to negative spillover effects. For example, multinationals may try to protect their firm-specific knowledge (see also Görg and Greenway 2004).

In Falki (2009), the relationship between the two variables is investigated for Pakistan, in the period 1980–2006. By using the production function based on the endogenous growth theory, the relationship between FDI and economic growth is analyzed. The study shows that a statistically significant relation between the GDP and FDI inflows in Pakistan is not found.

Temiz and Gökmen (2014) analyze the relationship between FDI and GDP growth in Turkey by using the Johansen cointegration test and Granger causality analysis. Their results suggest that no significant relation is determined between the FDI inflow and GDP growth in the country, both in short and in the long run.

Akinlo (2004) uses data for Nigeria in the period 1970–2001 to understand the relationship between FDI and economic growth. The ECM results show that both private capital and lagged foreign capital have small, and not statistically significant effect, on the economic growth. The results seem to support the argument that FDI in the extractive sector might not be growth enhancing as much as manufacturing FDI. In addition, the results show that exports, on the contrary, have a positive and statistically significant effect on growth; at the same time, labour force and human capital have significant positive effects on growth. These findings suggest the need for labour force expansion and education policies to raise the stock of human capital in the country, policies that are much more important for economic growth.

3 Model Specification

The dependent variable in our model is GDP per capita growth (GDPCG). We are interested to explore the role of global integration for promoting growth of economies. We use two measures (openness and foreign direct investment inflow) as our measure for global integration. Openness measures economy's trade as percentage of its GDP.

We have included, following the common literature (see Sect. 2), various control variables (inflation, gross capital formation, government expenditure, role of services sector), along with our variables of interest. The value added of this paper is to test whether global integration has a direct implication for GDPCG, or also have an indirect impact through other determinants of per capita income.

In our baseline model, we assume that GDPCG is dependent on several socio-economic variables in the country. Hence our baseline model:

$$\mathrm{GDPCG_{it}} = \alpha_{\mathrm{i}} + \beta_1 IGDPC_i + \beta_2 GI_{it} + \beta_3 Controls_{it} + \varepsilon_{it} \tag{1}$$

where i represents a country and t a time period. *IGDPC is an initial level of GDPC measured at the beginning of the period, GDPCG* denotes GDP per capita growth; *GI is our global integration variable, measured by openness* (trade as the percentage of the GDP) and FDI inflow (as the percentage of the GDP), controls represent a vector of control variables. Our control variables include gross capital formation, inflation rate, and government size measured as government expenditure as percentage of GDP. ε_{it} is an independently and identically normally distributed error term with zero mean and variance σ^2.

To check whether the impacts of other GDPCG determinants are conditional on global integration, we introduce an interaction term between openness and one of the variables of interest. Consequently, our model extends to

$$GDPCG_{it} = \beta_1 IGDPC_i + \beta_2 OPEN_{it} + \beta_3 Controls_{it} + \beta_4 \Psi_{it} + \beta_5 \Psi_{it} * OPEN_{it} + \varepsilon_{it} \quad (2)$$

where Ψ *is our variable for which we want to check its impact at different level of global integration.* $\Psi *$ *OPEN* is the interaction term of one of our variables of interest with openness. Since we are primarily interested in evaluating the effect of three explanatory variables and their interaction with openness, we will estimate four separate models extended to include:

1. OPEN * Value added of service sector;
2. OPEN * value added of industry;
3. OPEN * government expenditure;
4. OPEN and economic freedom index.

For example, to determine the impact of government size, measured as government expenditure as percentage of GDP, we estimate the model:

$$\mathrm{GDPCG}_{\mathrm{i,t}} = \alpha_{\mathrm{i}} + \beta_1 IGDPC_i + \beta_2 OPEN_{it} + \beta_3 Controls_{it} + \beta_4 GEXPG_{it} + \beta_5 GEXPG_{it} * OPEN_{it} + \varepsilon_{it} \quad (3)$$

As a next step, we calculate the marginal effect of each variable of interest at different levels of openness. For this purpose, we differentiate Eq. (2) with respect to that particular variable. Thus for the calculation of the marginal effect of government size conditional on openness, we take the derivative of Eq. (3) with respect to the financial development variable to obtain

$$\partial GDPCG / \partial GEXPG = \beta_4 + \beta_5 OPEN \quad (4)$$

The marginal effect of other variables of interest can be calculated accordingly.

4 Data Description and Analysis

We have employed the panel data estimation technique to analyze the determinants of GDP per capita growth. We have treated two independent variables, trade as percentage of GDP (openness) and FDI inflow as our main variables of interest. We consider both variables to represent the level of global integration in any economy. Through an empirical analysis we intend to understand the effect of global integration (measured as openness and FDI inflow) on per capita GDP growth. The period of analysis is 1980–2015. Our analysis focuses on emerging economies, BRICS and NEXT-11 (N-11 henceforth),[11] to understand the role of global integration in promoting growth in these economies.

Our dependent variable is GDP per capita growth. Data for dependent and explanatory variables are taken from World Development Indicators. Our explanatory variables include two indicators of Global Integration (GI), trade and foreign direct investment inflow as percentage of GDP. Other control variables include inflation, gross capital formation, and government expenditure, value added in industry and services sectors. Selection of our control variables is based on literature review presented in previous section. The precise definitions and data source of all variables used are given in Table 4 in the appendix. Summary statistics of the dependent variable and the independent variables are presented in Table 5 in the appendix.

The rationale for including the two key explanatory variables was discussed in Sect. 2.1. We repeat here the main arguments concerning FDI. FDI helps jumpstart an economy by adding to the direct capital financing as well as by being a source of valuable technology and know-how while fostering linkages with local firms. Lipsey and Sjoholm (2001) reviewed the literature and argued that there was evidence of positive effect of the spillovers from FDI. The impact of FDI inflow can also impact positively on human capital. When a Multi-National Company (MNC) invests in a country, it requires a certain level of human capital; in fact, MNCs usually invest in training the workforce. Therefore, the host country enjoys certain advantages such as knowledge accumulation, skill acquisition and manpower training. Moreover, MNCs can be advantageous to develop international networks to improve the trade potentials of a country. MNCs can also facilitate the movement of domestic goods and services across borders, create economies of scale and scope and thus contribute to economic growth (see Zhao and Du 2007).

[11]BRICS: Brazil, Russia, India, China, South Africa. N-11 excludes South Korea, because its growth is much higher as compared to other countries in N-11. We have however done empirical analysis including South Korea as well, but results remain more or less the same.

5 Empirical Results and Interpretation

The estimation results are reported in Table 1. The results are based on annual data over the period 1980–2014 for our group of emerging economies (BRICS and N-11). For better understanding the structural characteristics of these economies, Figs. 3 and 4 in appendix presents the level of global integration (openness) and GDP per capita in our sample countries.

Country specific random effects model is used. It is not feasible to use country specific fixed effects as we have time invariant variable (initial GDP per capita) as one of our explanatory variable. Since we are using time series data, serial correlation can be a potential issue. We applied panel data autocorrelation tests. The results suggest that there is not a problem of serial correlation in our residuals and in our time series macro variables; we can reject the null of unit root at 1 percent level of significance. Moreover, we also checked for possible multicollinearity among our explanatory variables; correlation matrix is presented in Table 6 in appendix.

5.1 Basic Model for Complete Sample (BRICS + N11)

In Table 1, our first column (model M1) is evaluating the impact of openness on per capita growth, which is positive and significant. The impact of other indicator (FDI inflow) is evaluated in model N1. Our hypothesis is that an increase in FDI inflow would lead to an increase in the GDP per capita growth. The results show that FDI inflow positively affects the dependent variable. The results are statistically significant. The coefficient values of FDI flow are higher as compared to coefficients of openness.

The results are in coherence with Ray (2012). In the study the author analyzed the casual relationship between economic growth and FDI in India. Study employed ordinary least square method and the results suggest that there is a positive relationship between FDI and gross domestic product.

We used Initial level of GDP per capita (INGDPC) in all specifications as a proxy for the level of economic development in the country. The coefficient sign is negative and significant which reflects the convergence in terms of growth among our sample countries.

In the second model (M2 and N2), gross capital formation growth is introduced as our control variable. The coefficient of GCFG is positive and statistically significant. Therefore, an increase in physical assets or the capital accumulation leads to an increase in the GDP per capita. Temiz and Gökmen (2014) believe that an increase in FDI flow by MNC can help a country to accumulate capital. FDI inflow is a substantial source in the case of low domestic savings rate and investment volume in a host country. Therefore, if the country's domestic saving rate is low then FDI might be used as a remedy to accumulate capital. Moreover, as proven in many empirical studies, an increase in GCFG would lead to an increase in GDP.

To control for biases and errors, we incorporated other explanatory variables: government expenditure growth and the values added to GDP by the Services and

Table 1 Impact of global integration on per capita growth—complete sample

	M1	M2	M3	M4	M5	N1	N2	N3	N4	N5
	b/se	b/se	b/se	b/se	b/se	b/se	b/se	b/se	b/se	b/se
INGDPC	−0.000*	−0.000***	−0.000**	−0.000***	−0.000***	−0.000*	−0.000***	−0.000***	−0.000***	−0.000***
	0	0	0	0	0	0	0	0	0	0
GCFG		0.172***	0.177***	0.174***	0.172***		0.164***	0.172***	0.165***	0.164***
		0.01	0.01	0.01	0.01		0.01	0.01	0.01	0.01
OPEN	0.032***	0.019**	0.020**	0.011	0.019**					
	0.01	0.01	0.01	0.01	0.01					
FDIFLOW						0.565***	0.300***	0.339***	0.298***	0.286**
						0.13	0.11	0.1	0.11	0.11
GEXPG			0.063***					0.062***		
			0.01					0.01		
INDEM				0.047*					0.059**	
				0.03					0.03	
SEREM					0.025					0.015
					0.02					0.02
Constant	1.793**	1.464**	1.032*	0.315	0.369	2.472***	1.972***	1.494***	−0.021	1.347
	0.9	0.6	0.58	0.95	1.14	0.7	0.46	0.43	1.01	1.11
Observations	509	494	489	494	494	485	470	465	470	470
Number of countries	15	15	15	15	15	15	15	15	15	15
R-Square	0.161	0.522	0.531	0.613	0.483	0.34	0.627	0.633	0.673	0.606
Chi-Square	12.772***	277.260***	511.387***	288.711***	278.009***	24.147***	256.954***	493.018***	268.592***	255.581***

Robust standard errors are reported below the coefficients. ***indicates significance at 1 percent level, **indicates significance at 5 percent and *indicates significance at 1 percent level

Industry sectors (GEXPG, INDEM, SEREM respectively): see the models M3 to M5 and N3 to N5 in Table 1). The result of the explanatory variables has their expected signs but are not statistically significant, except for government expenditure growth and value added to GDP by industry (see the models M3, N3 and N5). Government expenditure growth has the expected positive sign and is also statistically significant to 1%.

The sign and significance of our variables of interest (openness and FDI Inflow) remain positive and statistically significant in all specifications, which ensures the robustness and reliability of our results. Results reflect that global integration promotes economic growth. This result is coherent with De Mello (1999) study, among many others. De Mello in his investigation, conducted through time-series analysis, showed an increase in the amount of GDP through the effects of FDI on capital accumulation.

5.2 *Restricted Model (BRICS Only)*

To understand the potential differences between emerging economies of N-11 and BRICS, now we empirically investigate this relationship, impact of global integration on per capita income growth, specifically for BRICS group alone. Results are presented in Table 2.

The impact of openness and FDI inflow is positive and statistically significant (see models M1 to M5 and N1 to N5). The respective coefficient values for openness and FDI inflow are much higher as compared to the complete sample. This indicates the strong impact and significant role played by global integration in promoting economic growth in these economies. We can infer that in comparative terms trade opening (including FDI inflow) has been more beneficial to BRICS, that reached a comparatively higher level of development and were the first in the world (soon after the "Tigers" of South-Eastern Asia) to liberalize trade. On the contrary, it is likely that some countries of the N-11 group haven't yet reached the threshold level of development that allows them to fully exploit the benefits of trade liberalisation (as commented in the literature review section).

Coefficients of most control variables are significant and have the expected sign. Another interesting finding is that value added to GDP in services sector impact is negative and significant and the corresponding variable in the industry sector is positive and statistically significant. Industrial sector growth contributes positively towards global integration as economy's trade with other countries increases and it attracts more FDI in the country.

5.3 *Multiplicative Effects*

Our stated hypothesis is that global integration will impact GDP growth not only directly but also indirectly through other explanatory variables of growth. To

Table 2 Impact of global integration on per capita growth—BRICS

	M1	M2	M4	M5	M6	N1	N2	N4	N5	N6
	b/se	b/se	b/se	b/se	b/se	b/se	b/se	b/se	b/se	b/se
INGDPC	−0.001	−0.001**	−0.001***	−0.001***	−0.001***	−0.001*	−0.001	−0.001***	−0.001***	−0.000***
	0.001	0.00	0.00	0.00	0.00	0.00	0.00	0.00	0.00	0.00
GCFG		0.147***	0.146***	0.164***	0.156***		0.149***	0.145***	0.152***	0.154***
		0.014	0.018	0.018	0.017		0.015	0.017	0.016	0.017
OPEN	0.110***	0.088***	0.072***	0.032**	0.124***					
	0.016	0.013	0.013	0.014	0.014					
FDIFLOW						0.762***	0.635***	0.954***	0.633***	1.141***
						0.171	0.133	0.135	0.143	0.134
GEXPG			0.322***					0.249***		
			0.043					0.043		
INDEM				0.160***					0.154***	
				0.024					0.022	
SEREM					−0.192***					−0.118***
					0.023					0.02
Constant	1.022	0.504	−0.392	−2.588***	8.467***	4.155***	2.892**	1.233***	−1.923***	7.857***
	1.816	1.087	0.55	0.784	0.991	1.083	1.241	0.368	0.693	0.967
Observations	160	159	159	159	159	158	158	158	158	158
Number of countries	5	5	5	5	5	5	5	5	5	5
R-Square	0.289	0.421	0.781	0.747	0.872	0.645	0.684	0.891	0.862	0.965
Chi-Square	46.839***	181.192***	301.516***	274.793***	324.022***	24.059***	139.238***	335.313***	372.301***	337.970***

Robust standard errors are reported below the coefficients. *** indicates significance at 1 percent level, ** indicates significance at 5 percent and * indicates significance at 1 percent level evaluate this hypothesis, we introduce multiplicative terms of explanatory variables and openness. We introduce four interactive terms with our variable of interest. Empirical results are presented in Table 7 in appendix.

In the first model (SC1) we introduce an interaction term of openness with economic freedom index.[12] Similarly we include interactive terms between openness and government expenditure, openness and value added in services sector and openness and value added in industry (in models SC2 to SC4 respectively). The same analysis is presented for BRICS economies from model S1 to S4 in Table 7.

5.4 Marginal Effects

As suggested by Bramber et al. (2006), based on empirical results from SC1 to SC4, we calculated the respective marginal effects of interactive terms and present them in Fig. 1. Figure 1 presents the marginal effect of various explanatory variables on per capita growth, conditional on different levels of openness.

The solid line in each panel of Fig. 1 shows the marginal impact of one variable of interest on per capita growth at different levels of openness. The 95% confidence intervals around the solid line allow us to determine the conditions under which that variable has a statistically significant effect on GDP per capita growth labor productivity—that is, its effect is statistically significant positive (or negative) whenever the upper and lower bounds of the confidence interval are both above (or below) the zero line.

To evaluate and understand the difference between N-11 and BRICS, we have plotted marginal effect of these variables in Fig. 2 for BRICS only. Economic Freedom Index (EFI) is summary measure of economic freedom of the economy. While we found that economic freedom role is positive and statistically significant at low level of openness for the complete sample and then it turns negative, in case of BRICS sample results suggest that economic freedom role for promoting per capita growth increases with a higher level of global integration (see upper left panel in Figs. 1 and 2).

Similarly, we find that government expenditure impact turns negative after certain level of openness in the whole sample, however its impact remains positive and significant for per capita growth in BRICS economies and it increases with a high level of integration. This may be due to crowding out effect in developing and emerging economies of N-11 (see upper right panel in Figs. 1 and 2).

The marginal impact of value added in industry and services sectors at different levels of openness is presented in the lower panels of Figs. 1 and 2. We find that the

[12]The index measures the degree of economic freedom present in five major areas: (1) Size of Government; (2) Legal System and Security of Property Rights; (3) Sound Money; (4) Freedom to Trade internationally and (5) Regulation. It ranges from 0 to 10 (where 0 means no economic freedom and 10 means complete economic freedom).

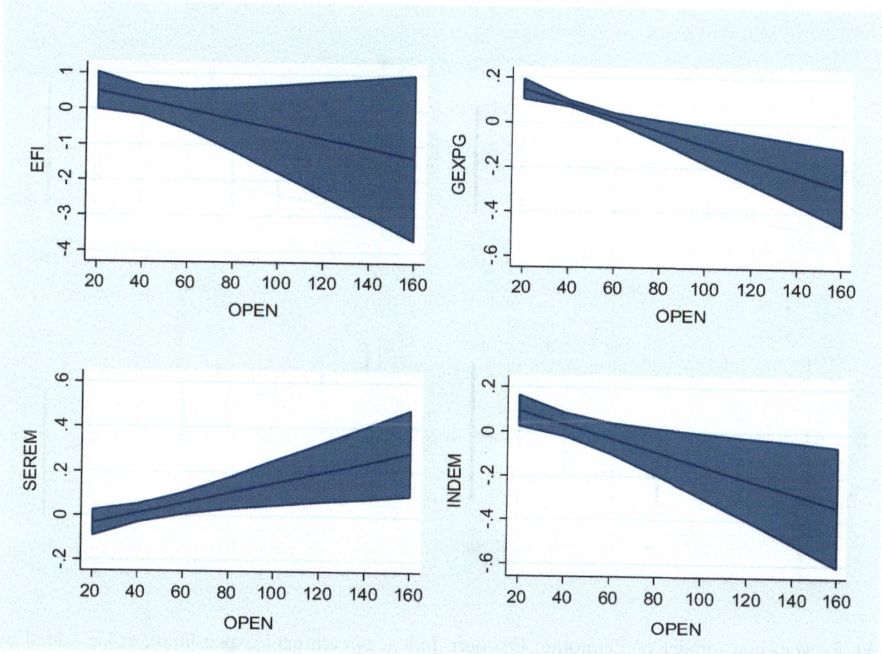

Fig. 1 Marginal impact of Economic Freedom Index, government expenditure, Value added by services and by industry on per capita income growth at different levels of openness—complete sample. This figure shows the impact of economic freedom index, government expenditure and value added by services and by industry on per capita growth at different levels of openness. These figures correspond to our main results as set out in columns SC1 to SC4 of Table 7. The upper panels show the marginal effect of economic freedom index (EFI) and government expenditure (GEXPG) and the lower panels show the marginal effect of value added by services sector (SEREM) and industry (INDEM) at different levels of openness

role of the services sector in promoting growth augments with an increase in global integration. This is true both for the complete and BRICS samples. It is statistically significant only in the case of BRICS economies. This might be due to the spillovers due to the trade. An increase in the trade would have positively affected the value added by the services sector.

On the contrary, the industry sector contribution towards promoting per capita income growth decreases with an increase in global integration. This finding is applicable both for the complete and BRICS samples. Sachs and Warner (1999) argued that extractive industries may have a negative effect on economy. The changes in local market structures, as a result of the incoming investment flows, could raise rent-seeking activity and deteriorate the institutions of the local economy.

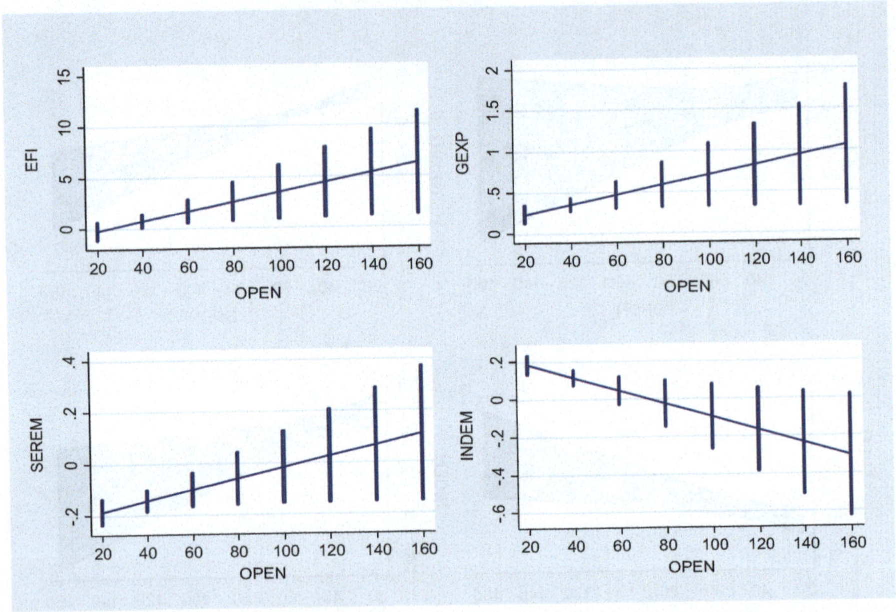

Fig. 2 Marginal impact of Economic Freedom Index, government expenditure, value added by services and by industry on per capita income growth at different levels of openness—BRICS. This figure shows the impact of economic freedom index, government expenditure and value added by services and by industry on per capita growth at different levels of openness. These figures correspond to our main results as set out in columns S1 to S4 of Table 7. The upper panels show the marginal effect of economic freedom index (EFI) and government expenditure (GEXPG) and the lower panels show the marginal effect of value added by services sector (SEREM) and industry (INDEM) at different levels of openness

5.5 *Robustness and Sensitivity Analysis*

Our empirical results suggest that global integration promotes economic growth. Now, we need to make sure that our results are consistent and robust. For this purpose, we conduct sensitivity analysis by changing our empirical model specifications and by treating for any potential endogeneity issue in our model.

There can be a potential issue of endogeneity and reverse causality, which can plague our results. High economic growth can lead to more FDI and trade openness. The FDI inflow is expected to supply technology, knowledge and capital, which lead to higher economic growth. In fact, the relationship between trade openness/FDI and GDP could be either way, i.e. trade openness/FDI increases GDP growth or GDP growth leads to trade openness/FDI increases.

In addition to what discussed in the literature review section, Basu et al. (2003) believe that a stable amount of GDP growth, capital accumulation, technology and competent workforce could foster foreign capital inflow that can make further contribution to the GDP growth, both in the short run and long run. Furthermore, high growth can lead to more gross capital formation, with the consequent need to import more machinery and capital goods.

To overcome this potential issue of reverse causality and check robustness of our findings, we employ the Hausman and Taylor (1981) estimation technique. This technique has been used extensively in the literature to control for endogeneity problems (see for example Degryse et al. 2012). Under this technique, weighted instrumental variable estimators are used to obtain coefficients by instrumental variable regression. Table 3 presents the estimation results regarding both the complete sample and the BRICS.

Results reconfirm that global integration has positive and significant impact on growth of the economies. Coefficients for openness and FDI inflow remain positive and statistically significant. The impact of other control variables also remains the same and they have the expected signs, which confirms the robustness of our findings. It also reflects that endogeneity issue does not contaminate our results.

In the next sensitivity test, we incorporate more control variables, which are potential determinants of per capita GDP growth. We introduce inflation, working age population, real interest rate, financial depth (measured as stock market capitalization), telephone line per 100 persons and literacy rate. Results are presented in the appendix in Table 8. Inflation rate (INF) impact is negative but not significant. High working age population share (TPOP) is positive for growth. Financial depth (FIND) impact is positive and significant. High real interest rate can lead to lower growth by increasing the cost of investment; in our results, RIR is positive but not significant. The impact of literacy rate (Lit) is positive for economic growth. The impact of trade openness (OPEN) remains positive and significant in five specifications; however, its impact is statistically insignificant in model M5. One potential reason for this can be the significant decline in the number of observations.[13]

Next we estimated our model by using fixed effects but excluding the initial GDP per capita from the explanatory variables. Our global integration variables remain positive and significant.[14] We also included the lag dependent varibale to check the persitance of this impact. We find that lagged dependent variable has positive and significant impact but global integration impact is still positive and significant. See Table 9. As our sensitivity tests show, our main conclusion holds: global integration as a positive role in promoting economic growth in our sample economies.

6 Conclusions

The impact of global integration on economic growth of N-11 and BRICS economies is confirmed in this paper. The impact of global integration—measured as foreign direct investment and share of foreign trade as percentage of GDP—on

[13]Time series data on literacy rate are not available.

[14]Results are not presented here but are available on request.

Table 3 Global Integration and Growth-Hausman Taylor Estimation

	Complete sample				BRICS			
	T1	T2	T3	T4	t1	t2	t3	t4
	b/se	b/se	b/se	b/se	b/se	b/se	b/se	b/se
INGDPC	−0.000*	−0.000**	−0.000***	−0.001***	−0.001*	−0.001**	−0.001**	−0.001*
	0	0	0	0	0.001	0	0	0
GCFG	0.133***	0.133***	0.130***	0.131***	0.146***	0.147***	0.151***	0.148***
	0.011	0.011	0.012	0.012	0.014	0.014	0.015	0.015
OPEN	0.043***	0.042***			0.093***	0.089***		
	0.01	0.01			0.013	0.017		
FDIFLOW			0.252**	0.212*			0.654***	0.515***
			0.109	0.112			0.133	0.152
SEREM		0.026		0.02		0.006		0.052*
		0.022		0.023		0.034		0.031
INDEM	0.012		0.067**		−0.047		0.091**	
	0.029		0.029		0.043		0.042	
Constant	0.41	−0.318	0.371	1.79	1.878	0.182	−0.131	0.565
	1.157	1.205	1.093	1.143	1.927	1.719	1.55	1.785
Observations	473	473	450	450	159	159	158	158
Number of Countries	15	15	15	15	5	5	5	5
Panel Standard Deviation	1.678	1.687	1.194	1.438	2.428	2.117	1.258	2.096
Wald Chi-squared	164.724	167.756	150.975	142.213	185.006	182.684	152.659	143.904

All models have been estimated by Hausman–Taylor regressions. Robust standard errors are reported below the coefficients. *** indicates significance at 1 percent level, ** indicates significance at 5 percent and * indicates significance at 1 percent level economic growth of per capita GDP is not only direct but also indirect, through various other determinants. Results with the multiplicative model confirmed our hypothesis. Our econometric results are robust, according to different specifications and tests.

However, the impact of trade openness (including FDI inflow) is much stronger and robust for BRICS compared to the whole sample of countries (N-11 and BRICS). It is likely that some N-11 countries, because of their comparatively lower level of development, cannot fully exploit the benefits of trade liberalization. They should, in the first place, invest in physical and human capital as well as improve their institutions, also learning from experiences of their peers, the earliest "globalizers" (China, India, and the other countries of the BRICS group).

Appendix

Table 4 Data description and source

Variables	Description	Source
GDPCG	GDP per capita growth	World Development Indicators (WDI)
OPEN	Trade as percentage age of GDP	WDI
FDIFlow	FDI Inflow (as percentage of GDP)	WDI
GCFG	Gross capital formation growth	WDI
GEXPG	Government final expenditure growth	WDI
INDEM	Industry value added (% of GDP)	WDI
SEREM	Services value added (% of GDP)	WDI
TELE	fixed telephone line subscription(per 100 people)	WDI
LIT	adult literacy rate 15+ both sexes percentage	WDI
INF	CPI	WDI
RIR	Real interest Rate	WDI
EFI	Economic freedom index	Fraser Institute

Table 5 Summary statistics

Variables	Mean	Std. Dev.	Min	Max
GDPG	3.044	3.878	−13.296	30.342
GCFG	6.909	12.158	−24.534	75.200
OPEN	46.999	24.915	12.009	169.535
GEXPG	5.512	26.758	−23.926	565.539
INF	40.847	214.321	−1.710	2947.733
TELE	8.003	8.887	0.103	38.981
IGDPC	1815.107	2154.165	0.000	6773.804
FDIFLOW	1.702	1.902	−2.757	11.939
SEREM	33.648	7.613	20.051	52.997
INDEM	49.193	10.589	19.736	71.030

Table 6 Correlation matrix

	GDPG	GCFG	OPEN	GEXPG	INF	TELE	IGDPC	FDIFLOW	SEREM	INDEM
GDPG	1.00									
GCFG	0.48	1.00								
OPEN	0.18	0.08	1.00							
GEXPG	0.42	−0.07	0.01	1.00						
INF	−0.10	−0.07	−0.16	−0.02	1.00					
TELE	0.04	0.03	0.08	−0.07	−0.01	1.00				
IGDPC	−0.29	−0.10	−0.16	−0.06	0.13	0.32	1.00			
FDIFLOW	0.14	0.09	0.51	0.03	−0.09	0.17	−0.08	1.00		
SEREM	0.04	0.03	0.35	0.06	0.14	0.24	0.12	0.13	1.00	
INDEM	−0.14	−0.09	−0.20	−0.16	0.00	0.43	0.55	−0.10	−0.44	1.00

Table 7 Global Integration and Growth: Multiplicative Model

	SC1	SC2	SC3	SC4	S1	S2	S3	S4
	b/se	b/se	b/se	b/se	b/se	b/se	b/se	b/se
OPEN	0.097	0.051***	−0.074*	0.138***	−0.206*	0.041**	0.013	0.159***
	0.06	0.009	0.039	0.043	0.118	0.02	0.053	0.046
EFI	0.779*				−1.108			
	0.41				0.795			
OPEN*EFI	−0.013				0.046**			
	0.01				0.02			
GCFG	0.126***	0.143***	0.135***	0.134***	0.137***	0.149***	0.140***	0.146***
	0.011	0.01	0.011	0.011	0.02	0.018	0.015	0.016
GEXPG	0.055***	0.209***			0.277***	0.12	0.257***	0.287***
	0.004	0.038			0.06	0.11	0.039	0.038
INGDPC	−0.000**	−0.000**	−0.000***	−0.000***	−0.001***	−0.001***	−0.001***	−0.001***
	0	0	0	0	0	0	0	0
OPEN*GEXPG		−0.003***				0.006**		
		0.001				0.003		
SEREM			−0.078*				−0.228***	
			0.042				0.043	
OPEN*SEREM			0.002***				0.002**	
			0.001				0.001	
INDEM				0.156***				0.254***
				0.054				0.049
OPEN*INDEM				−0.003***				−0.003***
				0.001				0.001
Constant	−2.377	−0.009	5.115***	−3.820**	6.415	0.646	9.124***	−7.313***
	2.489	0.683	1.978	1.915	4.287	0.754	1.972	1.577

(continued)

Table 7 (continued)

	SC1	SC2	SC3	SC4	S1	S2	S3	S4
	b/se	b/se	b/se	b/se	b/se	b/se	b/se	b/se
Observations	251	468	473	473	86	159	159	159
Countries	15	15	15	15	5	5	5	5
R-square	0.569	0.403	0.429	0.47	0.809	0.817	0.952	0.927
Chi-squared	306.636***	445.230***	177.433***	178.357***	180.427***	311.281***	461.183***	446.738***

Table 8 Sensitivity Analysis-Complete Sample

	M1	M2	M3	M4	M5	M6	M7	M8
	b/se	b/se	b/se	b/se	b/se	b/se	b/se	b/se
OPEN	0.037***	0.036***	0.031***	0.022*	0.003	0.002	0.016*	0.019**
	0.012	0.012	0.01	0.012	0.012	0.016	0.009	0.008
IGDPC	−0.000	−0.001**	−0.000	−0.001**	−0.001***	−0.000**		−0.000***
	0	0	0	0	0	0		0
INF	−0.002*							
	0.001							
RIR		0.018						
		0.019						
TPOP			0.000***					
			0					
GCFG					0.129***	0.204***	0.161***	0.179***
					0.016	0.022	0.009	0.01
TELE				0.072**				
				0.03				
EFI					0.874***			
					0.307			
Lit						0.042*		
						0.022		
FIND							0.016***	
							0.005	
GEXPG							0.060***	0.063***
							0.004	0.005
Total area								0.000*
								0

(continued)

Table 8 (continued)

	M1	M2	M3	M4	M5	M6	M7	M8
	b/se	b/se	b/se	b/se	b/se	b/se	b/se	b/se
Constant	1.213	1.856*	0.162	1.844**	−1.909	−1.014	0.292	0.661
	0.9	1.011	0.759	0.922	1.789	1.403	0.607	0.576
Observations	457	364	509	506	257	99	328	489
Number of countries	15	13	15	15	15	15	15	15
R-Square	0.034	0.035	0.026	0.032	0.227	0.511	0.648	0.513
Chi-Square	17.813***	14.283***	44.975***	17.942***	88.328***	93.274***	548.550***	518.438***

Table 9 Results with lagged dependent variable

	t1	t2	t3	t4
	b/se	b/se	b/se	b/se
GCFG	0.131***	0.127***	0.130***	0.131***
	0.012	0.013	0.012	0.013
INDEM	0.006		0.060**	
	0.029		0.028	
Lagged dependent variable	0.160***	0.138***	0.161***	0.152***
	0.041	0.044	0.043	0.046
OPEN	0.041***	0.039***		
	0.01	0.012		
INGDPC	0	−0.000*	−0.000***	−0.000***
	0	0	0	0
INF		0		0
		0.001		0.001
FDIFLOW			0.219**	0.167*
			0.109	0.1
Constant	0.019	0.185	−0.009	1.788***
	1.13	0.778	1.045	0.35
Observations	448	405	427	385
Number of countries	15	15	15	15
Panel standard deviation	1.51	1.104	0.99	0.403
Wald Chi-squared	187.799	151.687	182.135	165.283

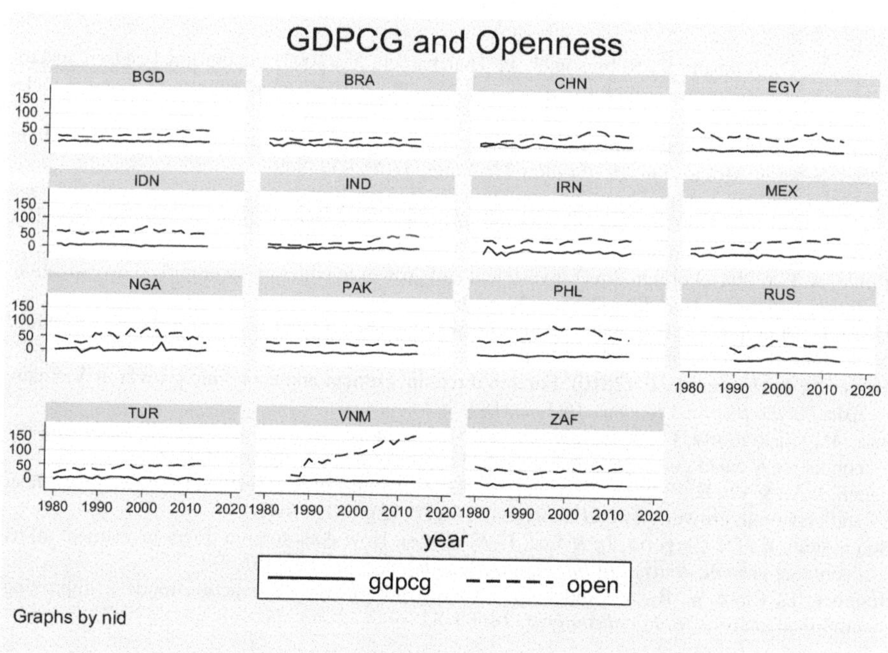

Fig. 3 GDP per capita growth and openness in sample countries

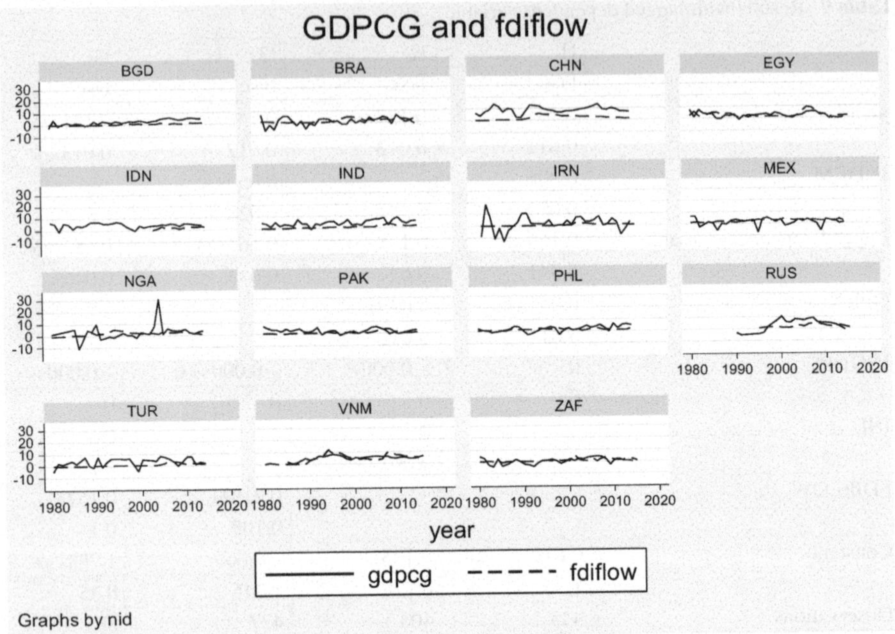

Fig. 4 GDP per capita growth and FDI inflow in sample countries

References

Acemoglu, D., Johnson, S., Robinson, J., & Thaicharoen, Y. (2003). Institutional causes, macroeconomic symptoms: Volatility, crises and growth. *Journal of Monetary Economics, 5*(1), 49–123.

Adams, S. (2009). Foreign direct investment, domestic investment, and economic growth in Sub-Saharan Africa. *Journal of Policy Modeling, 31*(6), 939–949.

Agrawal, P. (2005). Foreign direct investment in South Asia: Impact on economic growth and local investment. In *Multinationals and foreign investment in economic development* (pp. 94–118). UK: Palgrave Macmillan.

Akinlo, A. E. (2004). Foreign direct investment and growth in Nigeria: An empirical investigation. *Journal of Policy Modeling, 26*(5), 627–639.

Alfaro, L., et al. (2004). FDI and economic growth: the role of local financial markets. *Journal of International Economics, 64*(1), 89–112.

Anwar, S., & Nguyen, L. P. (2010). Foreign direct investment and economic growth in Vietnam. *Asia Pacific Business Review, 16*(1–2), 183–202.

Basu, P., Chakraborty, C., & Reagle, D. (2003). Liberalization, FDI, and growth in developing countries: A panel cointegration approach. *Economic Inquiry, 41*(3), 510–516.

Batten, J. A., & Vo, X. V. (2009). An analysis of the relationship between foreign direct investment and economic growth. *Applied Economics, 41*(13–15), 1621–1641.

Borensztein, E., De Gregorio, J., & Lee, J.-W. (1998). How does foreign direct investment affect economic growth. *Journal of International Economics, 45*, 115–135.

Bramber, T., Clark W. R., & Golder, M. (2006). Understanding interaction models: improving empirical analysis. *Political Analysis, 14*, 63–82.

Carkovic, M., & Ross, L. (2005). Does foreign direct investment accelerate economic growth. In *The impact of foreign direct investment on development: New measurements, new outcomes, new policy approaches.*

Chakraborty, C., & Nunnenkamp, P. (2008). Economic reforms, FDI, and economic growth in India: a sector level analysis. *World Development, 36*(7), 1192–1212.

Choe, J. I. (2003). Do foreign direct investment and gross domestic investment promote economic growth? *Review of Development Economics, 7*(1), 44–57.

Cieślik, A., & Tarsalewska, M. (2011). External openness and economic growth in developing countries. *Review of Development Economics, 15*(4), 729–744.

Degryse, H., Havrylchyk, O., Jurzyk, E., & Kozak, S. (2012). Foreign bank entry, credit allocation and lending rates in emerging markets: empirical evidence from Poland. *Journal of Banking Finance, 36*, 2949–2959.

De Mello, L. R. (1999). Foreign direct investment-led growth: evidence from time series and panel data. *Oxford Economic Papers, 51*(1), 133–151.

Dollar, D. (1992). Outward-oriented developing economies really do grow more rapidly: Evidence from 95 LDSs, 1976–85. *Economic Development and Cultural Change, 40*(3), 523–544.

Edwards, S. (1993). Openness, trade liberalization, and growth in developing countries. *Journal of Economic Literature, 31*(September 1993), 1358–1393.

Edwards, S. (1998). Openness, productivity and growth: What do we really know? *The Economic Journal, 108*(March), 383–398.

Falki, N. (2009). Impact of foreign direct investment on economic growth in Pakistan. *International Review of Business Research Papers, 5*(5), 110–120.

Frankel, J. A., & Romer, D. (1999). Does trade cause growth? *American Economic Review, 89*(2), 379–399.

Görg, H., & Greenaway, D. (2004). Much ado about nothing? Do domestic firms really benefit from foreign direct investment? *The World Bank Research Observer, 19*(2), 171–197.

Grossman, G. M., & Helpman, E. (1990). Comparative advantage and long-run growth. *American Economic Review, 80*, 796–815.

Harrison, A. (1996). Openness and growth: A time series, cross-country analysis for developing countries. *Journal of Development Economics, 48*, 419–447.

Hausman J. A., & Taylor W. E. (1981), Panel data and unobservable individual effects. *Econometrica, 49*(6), 1377–1398.

Herwartz, H., & Walle, Y. M. (2014). Openness and the finance-growth Nexus. *Journal of Banking and Finance, 48*, 235–247.

Huang, L. C., & Chang, S.-H. (2014). Revisit the Nexus of trade openness and GDP growth: Does the financial system matter? *Journal of International Trade and Economic Development, 23* (7–8), 1038–1058.

Iyke, B. N. (2017). Does trade openness matter for economic growth in the CEE countries? *Review of Economic Perspectives – Národohospodářský Obzor, 17*(1), 3–24.

Jadoon, A. K., Rashid, H. A., & Azeem, A. (2015). Trade liberalization, human capital and economic growth: Empirical evidence from selected Asian countries. *Pakistan Economic and Social Review, 53*(1 (Summer)), 113–132.

Khaliq, A., & Noy, I. (2007). Foreign direct investment and economic growth: Empirical evidence from sectoral data in Indonesia. *Journal of Economic Literature, 45*(1), 313–325.

Kim, D. H. (2011). Trade, growth and income. *Journal of International Trade and Economic Development, 20*(5), 677–670.

Kim, D., & Lin, S. (2009). Trade and growth at different stages of economic growth. *Journal of Development Studies, 45*, 1211.

Krueger, A. O. (1978). *Foreign trade regimes and economic development: Liberalization attempts and consequences. National Bureau of Economic Research*. Lexington, MA: Ballinger.

Lipsey, R. E., & Sjoholm, F. (2001). Foreign direct investment and wages in Indonesian manufacturing. In *NBER Working Paper 8299*. Cambridge, MA: National Bureau of Economic Research.

Liu, X., Burridgez, P., & Sinclair, P. J. N. (2002). Relationships between economic growth, foreign direct investment and trade: evidence from China. *Applied Economics, 34*, 1433–1440.

Makki, S. S., & Somwaru, A. (2004). Impact of foreign direct investment and trade on economic growth: Evidence from developing countries. *American Journal of Agriculture Economics, 86*, 795–801.

Marelli, E., & Signorelli, M. (Eds.). (2010). *Economic growth and structural features of transition*. London and New York: Palgrave Macmillan.

Marelli, E., & Signorelli, M. (2011). China and India: openness, trade and effects on economic growth. *The European Journal of Comparative Economics, 8*(1), 129–154.

Mercan, M., Gocer, I., Bulut, S., & Dam, M. (2013). The effect of openness on economic growth for BRIC-T countries: panel data analysis. *Eurasian Journal of Business and Economics, 6*(11), 1–14.

Milner, C., Vencappa, D., & Wright, P. (2007). Trade policy and productivity growth in Indian manufacturing. *The World Economy, 30*, 249–266.

Montalbano, P. (2011). Trade openness and developing countries' vulnerability: Concepts, misconceptions, and directions for research. *World Development, 39*(9), 1489–1502.

Mullen, J., & William, M. (2005). Foreign direct investment and regional economic performance. *Kyklos, 56*(4), 491–508.

Nannicini, T., & Billmeier, A. (2011). Economies in transition: How important is trade openness for growth? *Oxford Bulletin of Economics and Statistics, 73*(3), 287–314.

Narayanamoorthy, Sridharan, P., & Rao, K. (2009). Causal relationship between foreign direct investment and growth: Evidence from BRICS countries. *International Business Research, 2* (4), 198–203.

Nayyar, D. (2015). Globalization and Employment. *Indian Journal of Labour Economics, 58*(1), 87–97.

O'Neill, J. (2001). Building better global economic BRICs. Goldman Sachs, Global Economics, Paper No: 66: 1–16.

Pradhan, J. P. (2002). Foreign direct investment and economic growth in India: A production function analysis. *Indian Journal of Economics, 82*, 581–586.

Ramjerdi, H. P. (2007). Growth and productivity measures of China's due to international trade: PRC'S experience 1970–1993. *Asia Europe Journal, 5*, 253–265.

Ray, S. (2012). Impact of foreign direct investment on economic growth in India: A cointegration analysis. *Advances in Information Technology and Management, 2*(1), 187–201.

Rodríguez, F., & Rodrik, D. (1999). Trade policy and economic growth: A skeptic's guide to the cross-national evidence. *CEPR Discussion Paper, 2143*.

Rodríguez, F., & Rodrik, D. (2001). Trade policy and economic growth: A skeptic's guide to the cross-national evidence. In *NBER Macroeconomics Annual 2000* (Vol. 15, pp. 261–338). Cambridge, MA: MIT Press.

Rodrik, D. (2006). Goodbye Washington Consensus Hello Washington Confusion? *Journal of Economic Literature, 44*(4), 973–987.

Sachs, J. D., & Warner, A. M. (1995. revised 1997, 1999). *Natural resource abundance and economic growth* (National Bureau of Economic Research Working paper No. 5398), Cambridge, MA.

Sahoo, D., & Mathiyazhagan, M. K. (2008). Economic growth in India: Does foreign direct investment inflow matter? *The Singapore Economic Review, 48*(2), 151–171.

Sakyi, D., Villaverdeb, J., & Mazab, A. (2015). Trade openness, income levels, and economic growth: The case of developing countries, 1970–2009. *The Journal of International Trade & Economic Development, 24*(6), 860–882.

Sarkar, S. (2008). Trade openness and growth: Is there any link? *Journal of Economic Issues, 42*(3), 763–785.

Singh, T. (2010). Does international trade cause economic growth? A survey. *The World Economy, 33*(11), 1517–1564.

Squalli, J., & Wilson, K. (2011). A new measure of trade openness. *The World Economy, 34*(10), 1745–1770.

Stiglitz, J. E. (2002). *Globalization and its discontents*. New York: W.W. Norton.

Sulochani, R. S., & Leeb, K. (2015). Does openness lead to sustained economic growth? Export growth versus other variables as determinants of economic growth. *Journal of the Asia Pacific Economy, 20*(3), 345–368.

Sun, H., & Parikh, A. (2001). Exports, inward foreign direct investment (FDI) and regional economic growth in China. *Regional Studies, 35*(3), 187–196.

Tang, S., Selvanathan, E. A., & Selvanathan, S. (2008). Foreign direct investment, domestic investment and economic growth in China: A time series analysis. *The World Economy, 31* (10), 1292–1309.

Temiz, D., & Gökmen, A. (2014). FDI inflow as an international business operation by MNCs and economic growth: An empirical study on Turkey. *International Business Review, 23*(1), 145–154.

Tsen, W. H. (2006). Granger causality tests among openness to international trade, human capital accumulation and economic growth in China: 1952–1999. *International Economic Journal, 20* (3), 285–302.

World Investment Report. (2002). *Transnational corporation and export competitiveness*. UNTACD: New York and Geneva.

Xu, G., & Wang, R. (2007). The effect of foreign direct investment on domestic capital formation, trade, and economic growth in a transition economy: evidence from China. *Global Economy Journal, 7*, 1850107.

Yalta, A. Y. (2013). Revisiting the FDI-led growth hypothesis: The case of China. *Economic Modelling, 31*, 335–343.

Yao, S. (2006). On economic growth, FDI and exports in China. *Applied Economics, 38*(3), 339–351.

Zhao, C., & Du, J. (2007). Causality between FDI and economic growth in China. *The Chinese Economy, 40*(6), 68–82.

1948–2018: From the Free-Trade Vision to Protectionist Attitudes

Nicola Acocella

Abstract The institutions born at the end of WW II were inspired to a well-tempered liberism, but this principle has been disregarded in the following decades, especially since 1980, leaving instead room to an embittered liberism, which found its highest expression in the Washington Consensus. Then, the initial principle has largely been disregarded in practice. Even so, all in all, the Bretton Woods institutions, as they have evolved, have led to positive results. However, more recently populism and protectionism have spread, whose diffusion is the product, certainly excessive, of that exacerbate liberism, together with the blatant violations of the international rule performed by China. Political commentators and historians attribute the election of Donald Trump and its economic policy to the economic crisis begun in 2007–8, which was born rightly from such exacerbate liberism. The populistic waves that afflict Europe, herald of similar closures, can have similar foundations. If the world will stop in the path leading to commercial wars and closures of frontiers and will be able to reconstruct the season that led to the tempered liberism, is something desirable, but at present difficult to forecast.

Keywords Bretton Woods institutions · Exacerbate liberism · Populism · Protectionism · Washington Consensus

---

This article was awarded the 'Premio Economia Internazionale', in Genoa, Dec. 6th, 2018. The author is grateful to Amedeo Amato, the Jury of the Prize and the Genoa Chamber of Commerce. An earlier version of this chapter has been hosted on the website of The Institute for International Economics, Italy. This is an updated version.

N. Acocella (✉)
Memotef – Sapienza University of Rome, Rome, Italy
e-mail: nicola.acocella@uniroma1.it

L. Paganetto (ed.), *Capitalism, Global Change and Sustainable Development*, Springer Proceedings in Business and Economics,
https://doi.org/10.1007/978-3-030-46143-0_4

1 International Economic Policies: Institutional Foundations

The institutions that have governed international and national economic policies after the II World War and still now govern them were initially devised before the end of the conflict. In fact, from the Bretton Woods (USA) agreements of 1944 the International Bank for Reconstruction and Development (IBRD), usually called World Bank, and the International Monetary Fund (IMF) were born, as institutions aimed at favouring international cooperation in the economic, social and political field.

At least until 1971, keeping fixed—but adjustable—exchange rates around a parity was required by the Fund, even if some rules had to be obeyed tending to avoid frequent and diffuse changes in parities. These were admitted only to overcome deep and persistent imbalances of the balance of payments. To this end member countries could also implement some kind of capital movements control, since the Fund did not assist them to finance balance of payments imbalances due to capital exports. After the restoration of external convertibility in 1958, rapid shifts of capitals at an international level became possible and from the 1960s international capital movements increased. The IMF decided to finance them in order to protect the currencies that had been mostly hit, even if this was beyond its statutory obligations.

Also difficulties were faced in imposing re-equilibrating policies to both countries with persistent deficits and countries having persistent surpluses, even if the burden of adjustment was on both types of countries, as suggested by Keynes, who wanted to avoid a deflationary environment, which would have been the case if the burden had been imposed on deficit countries only.

In addition to the borrowing facilities available from the IMF to face temporary balance of payments imbalances, also long-term financing were (and still is) available to fund specific investment projects from the World Bank, at the usual banking conditions. The country asking for funding should obey some conditions, including measures restricting demand, market liberalization, etc. We will say more on that below, when we deal with the *Washington Consensus*.

The World Bank has, in fact, the mission to promote public and private investment in less developed countries and regions, with the aim to ease their living conditions. This task is performed by financing this investment and providing technical assistance for projecting and executing them and, more generally, for programming public action.

The Bank's guidelines in choosing the fields of investment to give priority to and financing conditions have changed over time. In the Nineties growth was favoured through free trade policies aiming at increasing the role of markets, as suggested by disputed interpretations of the growth and development in some countries, as those of South-East Asia. The countries recipient of funds were induced to 'adjust' their economies, in particular:

- by removing alleged obstacles and 'distorsions' in goods, labour and financial markets;
- privatizing large shares of the public sector of their economies and drastically reducing public deficit;
- liberalising goods and capital movements;
- adopting flexible exchange rates.

Specific insistence was laid on adjustment of internal conditions of markets to the international ones, i.e., on the integration of the economic system in the international context.

These measures go under the name of *Washington Consensus*, a term indicating an identical stance of the US Treasury, the Word Bank and the IMF, institutions all located in Washington.

According to many economists, the policies suggested or imposed by the World Bank and the IMF to some countries, such as Chile, Mexico, Turkey, East-European countries, South Korea, have often been too expensive, in terms not only of equity, but also of excessive and prolonged reduction in the growth rate of income and employment. Certainly, privatization and liberalization, often implemented in the absence of true markets (as in the case of East European countries), and the exposure, often abrupt, of economies not infrequently vulnerable to the conditioning and shocks deriving from international markets, particularly the financial markets, have produced situations of crisis that in some cases were heavy, raising the costs of adjustment. The policies advocated by the World Bank - and, to some extent, also those of the IMF—tending to favour privatization and liberalization have recently been questioned by their managing bodies.

In addition to the IMF and the World Bank, the institutions devised at the end of the II WW at Bretton Woods to implement international economic cooperation included also the *International Trade Organization*. This Organization, which aimed at ensuring cooperation in the field of trade policies, should have been instituted as an UN Agency, but this was never done, due to the opposition of the USA and other countries to the 1948 international agreement from which it derived. It was substituted by an agreement signed in Geneva in 1947 that should have been provisional, but has instead lasted until January 1, 1995, the *General Agreement on Tariffs and Trade* (GATT). This has eliminated quantitative restrictions such as quotas and other measures having a protectionist flavor and trade discriminations implemented by a country, e.g., through preferential tariff agreements with a foreign country. In addition, it has stabilized and progressively reduced tariffs, through subsequent sessions of multilateral treaties, the last one having been the *Uruguay Round*, closed in 1993.

Since January 1, 1995. the *World Trade Organization* (WTO) has begun to operate. It should be the place where multilateral trade negotiation and discussion of the different positions (also with reference to the implementation of the trade agreements stipulated in the past) take place. Then, it takes on the role of the organization devised at Bretton Woods. The WTO maintains the principles and many rules of the GATT, but has a more pronounced institutional feature, as it has a stable structure and foresees the almost automatic resolution of conflicts over trade between the member countries, which is performed by one of its branches, the *Dispute Settlement Body*.

Other international organizations, as the FAO, the ILO, the BIS, the OECD, play an important role for international cooperation. In addition, a number of regional organizations of countries, such as the EU—and, within it, the EMU—operate.

2 International Economic Policies and the Need for Coordination

2.1 The Inspiring Principles and Practical Applications

As to the principles inspiring the IMF, one can say that the institutions born at Bretton Woods were the expression of a tempered liberism, moderated by Keynesian thought.

The principle of free-trade was certainly the cornerstone that inspired the solutions adopted for the international setting. One can find it in the liberalization of exchanges of goods and in the multilateral approach we mentioned before. However, Keynesian thought implanted on this free-trade basis a series of public policies that tended prevailingly to ensure maintenance of full employment and coordination of economic policies. This thought was behind the symmetry required for policies tending to redress the balance of payments imbalances by both deficit and surplus countries. The deflationary policies enacted by the former in order to reduce their deficit would be less heavy, if balanced by policies of an opposing sign implemented by surplus countries.

For the same reason limits were imposed to free capital movements. In fact, complete freedom could translate in a further root of asymmetry, difficult to cure. In addition, similar arguments could be referred not only to international financial markets, but also—more generally—to all short-term financial markets.

For them Keynes's critique holds according to which speculation operates as in a beauty contest where wins he who designates not the prettiest faces, but who understands the prevailing preferences of the participants to the contest, thinking of their idea of beauty as to the face. Obviously, reasoning of a higher order of magnitude, of the kind 'everyone thinks of others thinking that, etc.' are possible, leading to indeterminate outcomes. More recently, the enormous size of international capital movements (with respect to the limited foreign exchange reserves of the various countries and the funds available through multilateral organisations), the high speed of their shifts, evidence of 'herd behaviour' (which takes place when most agents behave like others that they think more informed)—which replicates the mechanisms acting in a beauty contest—have increased the likelihood of self-fulfilling expectations having scarce or no relations with *fundamentals*, i.e., with the essential features of the economic systems in question.

As to the historical antecedents of Bretton Woods, confrontation between the results in terms of the GDP and international trade obtained as an effect of the liberalization in the period of the first globalization and those of the protectionist policies implemented in the interwar period was valuable. The former led to a remarkable increase in international exchanges, while the latter had the opposite effect.

2.2 *The Need for International Coordination*

For both the IMF and other international institutions (in particular, those dealing with the barriers that may be implemented by the various countries) the need for international coordination derives from the necessity to avoid 'races to the bottom' by the various countries.

In a world where, on the one hand, the competitive pressure has increased as an effect of transportation and communication progress and, on the other, the action of international organisations has succeeded in lowering the classical tariff and non-tariff barriers, a country can still strengthen its competitive position by exploiting its more permissive policies in terms of environmental, antitrust and other policies vis-à-vis foreign direct investment and transnational enterprises, social protection and tax policies.

For instance, in order to attract foreign investment, it can lower taxes applying to foreign companies, inducing other countries to do the same. More generally, similar attitudes can be taken for any possible object of regulation, especially in the case when a country suffers from backward conditions. A country can thus try at least to gain benefits in the short-run, even if to the detriment of itself and other countries in the long-run. In fact, it is clear that the permissive policies enacted by some countries can be offset by similar policies of other countries, tending to re-establish the initial conditions. The race to the bottom between national governments, while not solving the problems of more backward countries, would risk to downgrade public policies to levels and modes that limit an effective control of markets or their substitution.

The need for international coordination is clear. However, we must notice right now the limits of an international coordination of national economic policies that neglects the equity aspects in the international distribution of income and wealth, not only as a source of 'distortions', but also as true distortions to remove.

Safeguard clauses and the exceptions granted to LDCs in the implementation of the rule agreed on in the international fora recognize this specific status of such countries, which is necessary, even if certainly not sufficient to allow them to increase their production, which needs complementary projects to strengthen the exporting capacity of LDCs to which various international public institutions have cooperated.

There are then two reasons underlying the policy attitudes worth of specific attention by the international institutions entrusted with supervision of international trade: the competitive pressure (together with the joint need to avoid a race to the bottom) and the state of backwardness of the countries that often engage in the policies mentioned.

3 The Outcomes of Existing Institutions

3.1 A General Picture

A general picture of the indicators relative to the various aspects of the second globalization that allows us to compare it with the first globalization is offered by Table 1, which, however, stops at 2007 and thus neglects the effects of the economic crisis.

In 2015 and 2016, international trade has suffered the negative consequences of the crisis, more than the income produced in the various countries, as it happened in 2009, but already in 2018 it has come back to growth rates higher than those of income. Trade expansion in many cases leads to situations of imbalances: exports rise more than imports in some countries, determining a surplus in the current account of those countries, while the opposite happens in other countries.

As to the sectors originating exports, there has been a strong growth of exports of services, whereas exports of both agricultural and, even if to an even more reduced rate, of manufacturing have risen less.

In the most recent years there has been a further shift towards international exchanges relative to the information industry, a term indicating all productive branches having a high informative content, e.g., those referring to the computer industry, information programming, telecommunications, design of systems. In this realm, the spillover effects deriving from these exchanges for the economic systems of emerging markets have a specific interest. China has been particularly able in

Table 1 Indicators of globalisation in the XIX and XX century. % changes, except otherwise indicated (Source: WTO 2009)

World	1850–1913	1950–2007	1950–73	1974–2007
Population growth	0.8[a]	1.7	1.9	1.6
Total (real) GDP growth	2.1[a]	3.8	5.1	2.9
Per capita (real) GDP growth	1.3[a]	2.0	3.1	1.2
Growth of trade (in real terms)	3.8	6.2	8.2	5.0
(Net cumulative) migration to US, Canada, Australia, NZ				
Millions	17.9[a]	50.1	12.7	37.4
% annual rates of changes of migrants	0.42[a]	0.90	0.55	1.17
Year			1982	2006
FDI stock, in % of world GDP	–	–	5.2	25.3

[a]1870–1913

Table 2 Stock of international migrants as a percentage of local population, 1990, 2017 (Source: United Nations 2017)

	1990	2017
World	**2.9**	**3.4**
More developed Regions	**7.2**	**11.6**
Less developed Regions	**1.7**	**1.8**
Austria	10.3	19.0
France	10.4	12.2
Germany	7.5	14.8
Greece	6.0	10.9
Italy	2.5	10.0
Poland	3.0	1.7
United Kingdom	6.4	13.4
Spain	2.1	12.8
Sweden	9.2	17.6
Hungary	3.3	5.2
Australia	23.2	28.8
United States of America	9.2	15.3

exploiting purchases of the information industry products for its industrial development policies.

The rise of movements of people has been lower with respect to the high rates of growth of goods and capital exchanges. If we relate these exchanges of people to the world population, the ratio remains relatively constant over the years, even if it shows an increase in the recent years. It is however true that even constancy of this ratio implies a substantial absolute increase, since the world population has significantly risen, even if at a lesser pace than in the early Sixties of the Twentieth century (around 1.1% now, against 2.2% of those years).

The population of developed countries has increased or remained stationary, due to immigration. For the world as a whole, it appears that the number of migrants has barely risen, but this can largely be attributed to the rapid growth of the population of the countries of origin of migrants, as shown by the following Table 2:

3.2 *Specialisation Opportunities for Some Countries in Segments of Chain Values*

An important characteristic of international trade in the last decades has been the growth of the specialisation of various countries in different segments of the same sectors, which originates what is called *intra-industry trade*.

The underlying operation is the breakdown in value chains. For instance, production of a vehicle can be broken down in that of motors, gears, wheels, doors, windows, etc. The production of each segments of the final product can be localized in a different country, on the basis of the characters of its production, on the one hand, and those of the various countries, in terms of industrial structure, labour skills and costs, on the other. For instance, door production can require a less specialised labour force than that of motors. In this case, the first can be localized in countries relatively less developed than the latter.

This is of specific importance, as it opens the way to produce the different component parts of a car in countries that in other cases could have never produced whole cars.

3.3 *The Reduction in Inequalities Between Countries, Together with Their Rise in Some Developed Countries, Mainly Anglo-Saxon Countries*

Inequality is a multidimensional phenomenon, which can take different forms and can be measured by different 'spaces': consumption, income, wealth, utility, capability and opportunities of choice. The most common spaces are income or wealth. With respect to the first space, there are different dimensions of inequality: geographical, personal or factor distribution. Wealth inequality can refer to total wealth—net of debt—or to specific categories of wealth (financial assets or houses and land). For each dimension, different indicators can be used, as, e.g., the income pertaining to the richest (or poorest) 1% (or 10% or another share) of the population. Obviously, these indicators give information only for a share of the population. For example, if we know that the income share of the richest 1% of households rises, we ignore what happens to the remaining part of the population, i.e. whether the other 99% have a uniform loss of their share or it is the middle class that loses, with some possible increase of the share of the others. Synthetic measures of the distribution exist, such as the Gini index (or coefficient), but their complexity does not imply that their indication is completely truthful.

Measurement of inequality is in any case preliminary with respect to any other analysis and to the policy suggestions aiming at its reduction. Existence of many indicators should not discourage their use. In fact, by referring to multiple indicators one can enrich the analysis and better calibrate public intervention.

As to the indicators of geographic inequality, they can be referred to the different regions, or provinces, of a given country or to the whole world.

At the world level, inequality can be measured differently, since measurement is more complex than for a country, not only because it is difficult to have the wealth of information available at a national level. Only a few sources give information of the income of the richest or the poorest 1% (or 10% or any other share of the world population), which in this case can come from the different countries, or similar kinds of information. Usually, one can have a synthetic index of world inequality taking for each country its average income, which can be weighted or not by its population, as this index is easier to calculate than the distribution of income for the

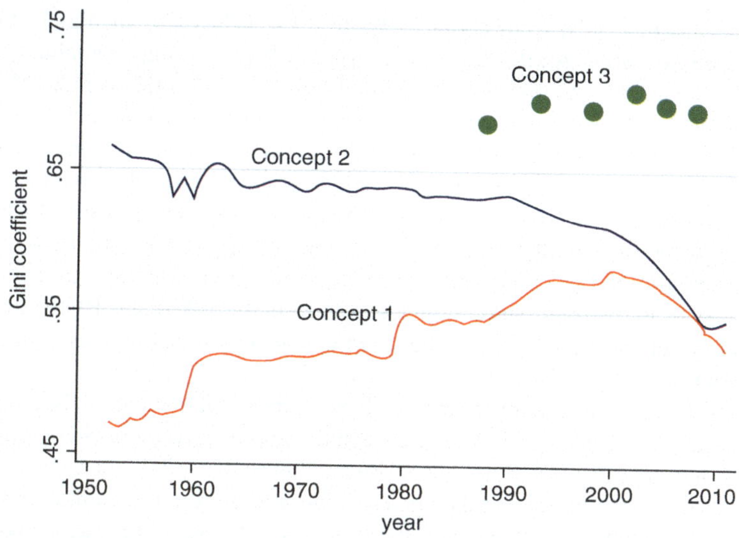

Fig. 1 International and global inequality, 1952–2011 (Source: Milanovic 2013)

whole world population calculated in a way similar to that of a single country. In fact, only a few sources offer such an index of world inequality. In the case we will show the index used is Gini coefficient.

Figure 1 gives an idea of the three different indicators, i.e., international inequality considering the average income of each country, weighted and unweighted for the population of each country, and world inequality.

The different indexes of income concentration show different time trends. The first one, which considers the average income of each country, rises almost continually up to 2000 and then lowers until, approximately, 2008. The increase in the period 1980–2000 is due to the fact that in this period the average income of developed countries rises more than that of other countries (Remember that a rise in the value of the Gini coefficient indicates a worsening of inequality.) The index calculated by weighting a country by its population (Concept 2) practically declines over the whole period. This depends on the fact that the lower income of the LDCs appears as higher as an effect of the increase in the population of these countries. In other terms, with this index poorer countries have a higher weight, which attenuates the inequality and thus lowers the value of the coefficient.

The third index, i.e. global inequality, as measured by the global Gini coefficient, is available only for the two decades after 1988 and oscillates from one year to the next.

As to income inequality within specific countries, the findings of various studies, referring to different periods, countries and indicators, differ. From the end of the Civil War to the start of World War II, the United States shows a profile of rising personal income inequality in terms of the Gini coefficient, mainly as an effect of rising monopolisation. From 1976 to 2012 the share of the richest 1% of the population has increased from about 8–9% to 20%. It has also increased, but at a reduced rate in other Anglo-Saxon countries, whereas it has remained almost constant in Japan. The picture for Continental European countries is rather different from that of Anglo-Saxon countries. A general feature of Continental European countries is that after the II WW inequality has first lowered and then remained constant even after the Great Recession, mainly as an effect of the operation of the welfare state. Elsewhere, some reduction of inequality has taken place in the last two decades in some (not in all) South American countries, partly as an effect of lower interest rates and higher commodity prices, and in some cases due to a retreat in many places from the free-market policies enacted in the 1980s and 1990s. In China, India, Indonesia and South Africa, the share of the richest has increased in the last 3–4 decades.

Various factors have contributed to these trends. Globalization has certainly played a role: on the one hand, the increase in the goods movements has contributed to reduction of inequalities; on the other, liberalization of capital and financialisation have increased them, by raising the shares of income accruing to capital and giving rise also to financial crises. However, the differences between the various countries cannot be explained without referring also to their domestic policies, having to do with the different degree of liberalization for labour, goods and capital markets and to the income and wealth tax policies, e.g., inheritance taxes.

3.4 Financial Crises

The most relevant issue recently arisen is that of the financial and economic crisis begun in 2007, which has been named the Great Recession, due to its significance. In fact, in the last decade or so, most advanced countries have suffered the deepest economic crisis after 1929 (the Great Crisis), a phenomenon—the crisis—that seemed to be confined to history books, remaining outside reality. The Great Crisis lasted long, originating a kind of stagnation, which was exited at the end of the Thirties only as an effect of the strong increase in public expenditure tied to the impending war. After WW II there were various cases of crisis, which, however, have been limited to a few countries (often developing countries). In the USA the dot-com crisis arose in 2001 and still in 2018 crises have arisen in Turkey and Argentina. The crisis of 2007 has instead concerned mainly developed countries and has been by far the deepest crisis after WW II, recalling not only the intensity but also the sad episodes (like suicides, which have risen by 10,000 from 2008 to 2010, according to some estimations) of the Great Crisis.

It arose in the USA and transmitted immediately to Europe. Its proximate roots are in the accumulation of private capital and the improper forms it had taken as well as the speculative activities that preceded it. These recall to some extent the episodes of the recent and remote past.

In 2009 the crisis had begun to recede in the USA, but became deeper in Europe, transmitting itself to the sovereign debt, i.e., to public debt. The reason for that should be searched in the improper character of both the European institutions and policies. It is true that a similar problem arises also in the USA, where the crisis had been nourished by financial deregulation. However, the fiscal union existing in the USA has permitted to avoid there the indicated transmission, i.e., the transmission of the crisis from the private to the public sector, which has instead happened in the EMU, where there is no fiscal union. In 2009, the last common year of the crisis in the USA and Europe, the GDP has fallen by 3.1& in the USA and 4.4% in the Euroarea.

To have an idea of the effects of the crisis, it is indicative that only in Italy one million jobs have been lost and 82,000 firms have gone bankrupt in the period 2008–2015. In the EMU the rate of unemployment has risen by almost 2.5 p.p. in the 8 years after 2007, reaching the value of about 1/7 of the activity rate (i.e., 1/7 of the total of employed and unemployed), but it has more than doubled in Spain, reaching the value of 23%. On the whole, the effect on the unemployment rate ofOECD countries has not been so high, reaching the highest value of 8.4% in 2009 and then lowering to 7.7of the total of employed and unemployed), but it has more than doubled in Spain, reaching the value of 23%. On the whole, the effect on the unemployment rate of OECD countries has not been so high, reaching the highest value of 8.4% in 2009 and then lowering to 7.7% in 2013 and 6.4% in 2016. In the non-OECD countries the impact has been relatively lower: the NICs (*Newly industrialising countries*), which include BRICS—i.e., Brazil, Russia, India, China and South Africa—and other lower-size countries) were hit by a maximum rise in the unemployment rate of 4 p.p. in South Africa and of 2.4 p.p. in China. In some cases the rate of unemployment underestimates the negative effect of the crisis. In fact, 'discouraged' workers are induced not to offer themselves on the labour market, as they consider unlikely being hired. Among them there are the so-called NEET, acronym of *Not (engaged) in Education, Employment or Training*, whose number has risen significantly.

3.5 The Lessons that Have Derived

The recent financial and exchange crises have offered a number of lessons that can be useful for domestic and international policy. They refer, in particular, to:

1. the need of adopting more effective policies to supervise domestic credit and financial markets, by a more extensive implementation of the *standards* indicated by the Basle Committee of the BIS for bank supervision referring to bank capital, prudential control for systemic risk and information requirements to customers;
2. the cases when adoption of fixed exchange rates is not advisable; in fact, even if useful to acquire credibility in the initial phases after a period of high inflation, fixed exchange rates are sustainable with difficulties by countries characterized by real or financial weakness and, as for the EMU, without a fiscal union;
3. the effects of international capital liberalisation in originating the crisis and the need of controlling them someway;
4. the role of international organisations, in particular the IMF.

4 Critiques to International Organisations

4.1 Critiques to the IMF. The Hegemony of the United States and the Indiscriminate Applications of Liberism

The IMF has largely intervened to sustain the countries experiencing exchange and financial crises and has provided hundreds of billion dollars. In order to have more funds available the Fund has indebted to some countries and decided to rise quotas. For each of the countries in crisis the Fund has done programmes asking for undertaking of specific engagements (in particular, adoption of restrictive budget or monetary measures), in exchange for the credit granted (credit conditionality).

The conduct of international institutions, in particular of the IMF in exchange crises has been the object of numerous critiques by economists, politicians and other people in charge of public institutions. They refer to various aspects of this conduct.

1. Interventions of an international lender of last resort, as some think the IMF is, reduces the ability of self-assessment and discipline, accentuating the moral hazard of the various economic agents and countries, each of which engages in riskier actions, precisely as a result of its conviction that intervention by the Fund will avoid its bankrupt. As a matter of fact, one can notice that the likelihood of moral hazard is highly attenuated by the following circumstances:
   (a) the intervention does not eliminate any damage for the borrowers; it certainly avoids its insolvency, but does not cancel the negative consequences, in terms of reputation, deriving from the inappropriate conduct that has made necessary the intervention;
   (b) in any case, the Fund's intervention contains other penalties for the debtor whose insolvency is avoided; in particular, the interest rates applied by the Fund are higher than those of the market and, above all, as a rule the Fund does not bear all the losses of the debtor country. Providing liquidity to it has the only purpose of avoiding propagation of the financial crisis, not to eliminate some 'therapeutic' effects of the crisis.
2. A second critique concerns scarcity of the funds provided by the Fund. This critique is symmetric to the previous one. However, refusal of the argument according to which the role of lender of last resort reduces self-discipline of economic agents does not necessarily imply that one asks a rise in funds provided by the IMF. In fact, some paradoxical aspects in the behaviour of markets is that their reaction to the Fund's intervention have been more negative the higher the size of the credit granted. A similar reaction has to be traced back once again to the circumstance of the high information asymmetry existing in financial markets, which makes any intervention of international authorities to be interpreted as a signal of crisis so much deeper the higher the size of the funds lent.
3. A third reason of critique has to do with the conditions imposed on the countries financed by the Fund, which can deepen the crisis. For instance, one such

condition is restriction of aggregate demand, which is often required to facilitate the adjustment of goods movements. This restriction limits the extent of the devaluation needed to rebalance them, but can cause recession and a rise in unemployment, aggravating in the short run financial problems, reducing solvability of banks and the service of public debt.

4. A fourth reason of critique to the action of the IMF has concerned the policies of liberalisation, privatisation and deregulation imposed together with deflationary measures within the *Washington Consensus.*
5. An aspect to discuss refers to full liberalization of capital movements. We have already said that it can play a negative role for the stability of the system. Similar considerations had inspired the IMF statute, which allowed countries' control on capital movements. This should be left to national sovereignty, consistently with Keynes's indications at Bretton Woods. The rule has never been cancelled and holds still now, even if it has fallen into disuse due to the implementation of the Washington Consensus. In fact, at the end of the Nineties the opportunity to change the statute—by introducing the obligation to capital movement liberalization was discussed—but the financial crises that had recently occurred, such as those of South-East Asian countries, and some studies within the Fund, led to an opposite decision. In addition, the Fund's intervention to sustain the imbalances due to capital movements, rather than to the current account, shows the inadequacy of its action and violation of its statute.

In acceptance of some of the critiques addressed to it, after 1999 the IMF has devised plans in order:

1. to ensure higher transparency of the conditions and decisions of the various countries, in addition to the Fund's orientations and decisions;
2. to strengthen the financial sectors of the various countries through better rules and supervision and closer international coordination;
3. to increase the activity of prevention of, and response to financial crises, in particular by introducing a new credit line to use for prevention;
4. to limit the conditions imposed on the countries that are beneficiary of financial assistance only to the measures that are strictly necessary for macroeconomic stabilization and rebalance of foreign accounts.

In addition, recently the Research Department of the IMF has recognized the opportunity to control capital movements in emerging and developing countries, having verified that the control has tended to induce a less risky and vulnerable structure of incoming capital.

With reference in particular to the action played by the United States in the realm of international payments, neglect by this country of the balance of its external payments has allowed to ensure the equilibrium of the whole system of international payments. One should think that the total of the balances of the foreign accounts of all countries cannot be but zero. In a two-country world, if country A has a surplus of 50, country B cannot but have a deficit of the same amount. Similarly, if—as it happened in reality—all the other $n-1$ countries wanted to have a surplus or a balance equal to zero, the United States should carry a deficit. This happened since WW II to the second half of the Fifties, but is only one aspect of the role of the USA in the international system of payments, which hides the fact that, by not caring for the balance of foreign accounts and suffering in practice a deficit, the United States enjoyed the condition of 'living in excess of their possibilities' or, more precisely, enjoyed imports higher than exports and freely exported capital abroad in order to buy firms or to lend for economic and political purposes.

Ultimately, having continuous deficits in one's balance of payments implies paying foreign exchange outflows in excess of inflows by its own currency. This explains why the dollar acquired a hegemonic role in the international monetary system. However, the establishment of the dollar's supremacy was due—in addition to the continuous balance of payments of the USA—also to the low impulse given to the diffusion of the currency devised in the mid-Sixties, i.e., the Special Drawing Rights, a purely accounting currency, which should have complemented the dollar as an international currency.

The last aspect of the United States' hegemony in the action of the Fund has to do with the influence of this country—also as a reflection of the positions and the interests of the financial environment—in the analyses and choices of this institution, as in the case of the report on the economic miracle of the South-East Asian countries, in which the IMF presented an enhanced pro-market interpretation of the determinants of their growth, whose validity has been questioned.

The United States have influenced not only the action of the IMF, but also that of the World Bank, as should be clear from what we said about the Washington Consensus.

4.2 *Speculative Capital Movements and Foreign Exchange Crises*

Numerous foreign exchange and financial crises have marked the events following WW II. In particular, they have been numerous after the Eighties, in relation with the restrictive monetary policy inaugurated by the United States in 1979, which caused marked increases in real interest rates, leading to the dollar appreciation that aggravated the service of debt for some Latin-American countries, thus requiring higher outflows of foreign exchange from them. Then, in the Nineties, a crisis of South-East Asian countries and Russia took place. The recent financial crisis, which has shaken the foundations of the European Monetary Union, with the possible exit of some country, has not ended the series, which has continued in 2018 and later, as said with the crises of Turkey and Argentina. Free movements of massive amounts of capital are a factor that undoubtedly eases and amplifies the diffusion of foreign exchange and financial crises.

All the countries involved in foreign exchange crises had previously liberalized international capital movements and actually a study of the *Federal Reserve Board* showed that already towards mid-Nineties the incidence of bank crises was highly correlated with the degree of capital market liberalization. One should also say that, after the inception of the crisis, various countries tried to re-introduce controls over capital movements, which however were ineffective. This does not demonstrate that controls are worthless, but, by contrast, reinforces the idea that a country should not prematurely liberalise capital movements. In fact, under asymmetric information in capital markets, a premature liberalization can lead to a considerable inflow of capital, which however rapidly flow out of the country, after reintroduction of controls, as this is interpreted as a signal of the heaviness of the crisis. In other terms, the issue of outflows of capital is strictly tied to the rules applying to incoming flows and foreign exchange crises cannot be avoided only by the introduction of a prohibition of capital outflows. If the issue is dealt with only from this point of view, it seems unsolvable, due to the indicated signaling effects. If, instead, rules introduce a selection of incoming capitals, the negative effects are absent or will be attenuated. Entry regulation, which can substitute Tobin's tax, can tend to discourage short-term inflows, e.g., by introducing some additional reserve requirement for short-term deposits coming from abroad, as done in the past in Chile and Colombia.

The argument usually raised in favour of capital movement liberalization is that this generates efficiency. It is rightly noted that this liberalization can have efficient effects complementary to those deriving from international trade liberalization. In fact, while the latter allows exchange of a good on terms more advantageous than those offered in the domestic market, international capital liberalization allows outlets or provisions more advantageous than domestically, in an intertemporal setting, which is where capital transactions should be placed. Liberalisation of capital movements should allow capital to move from the countries where it is abundant (and the rate of interest is low)—which are usually developed countries—to countries where it is scarce (and the rate of interest is high), usually LDCs. It is said, in particular, that capital movement liberalisation benefits LDCs for three reasons, due, first, to the direct effects that capital inflows have on the possibility to carry out investment and, therefore, to raise their growth rate. In fact, the higher capital inflows that are thought to be related to the liberalization should allow higher net imports, without threatening the balance of payment equilibrium. LDCs should then have the possibility of investment exceeding the low level of domestic saving. In addition to these direct benefits, the international mobility of capital should offer indirect benefits, e.g., as they offer the possibility to share risks internationally. Finally, the international mobility of capital should allow, in particular, to redirect the destination of resources from countries having an 'incorrect' conduct and bad performance (e.g., having low growth, high inflation, balance of payment deficit) to those more 'virtuous' and with sound fundamentals (with fast growth, low inflation and balance of payments, net of short term capital, tending to equilibrium).

However, these arguments should be the object of a number of critiques, referred in particular to short-term capital movements (and then to short-term international financial markets) and, more generally to the working of short-term financial markets, also in a closed economy.

1. First, Keynes's critique to the action of speculation (beauty contest) holds.
2. The enormous size of international capital movements (with respect to the limited foreign exchange reserves available by the various countries and the funds that can be obtained from multilateral organisations), the high speed of movements and the occurrence of herd behaviour raise the likelihood of self-realising expectations, having scarce or no relationship with fundamentals.
3. With reference to asymmetric information, financial liberalisation, by leading—at least in the short run—to a rise in competition, reduces the profit margins of financial intermediaries, inducing them to raise their activity in completely new fields, possibly less known (with a possibility of a higher adverse selection) and riskier (moral hazard).
4. Capital movement liberalization has increased the volatility (i.e., the erratic variations of prices) and, therefore, the uncertainty of financial and foreign exchange markets. The higher volatility of financial markets has increased the search for short-term positions, which are less uncertain, but these positions are more affected by signals, news and contingent feelings, which can increase volatility.
5. Short-term capital movements imply an excessive variability of exchange rates that hinders international trade. In addition, they can generate misalignment of exchange rates, i.e., the tendency of these rates to remain at values that are continually above or below a value of them that ensures the 'fundamental' equilibrium of the balance of payments, with ensuing costs for the real economy.
6. With reference to the ability of international capital movements to determine a 'virtuous' conduct and outcomes for the various countries, one should notice that, for the reasons already indicated, financial markets tend to determine reactions that are excessive with respect to those that may be required to adjust the economic position of a country, worsening its position and, in case, raising its fragility.

5 National Policies

Until the end of the Sixties also national policies were inspired to a tempered liberism, similarly to international economic policies. Possibly, national policies were more sensitive to state intervention. In fact, Western countries enacted policies tending to favour employment and growth, controlling prices and incomes (in Anglo-Saxon countries, this was a continuation of policies implemented during WW II), redistributing incomes by taxes and public expenditures, and a large sector of public enterprises and in some cases adopted national planning.

The theoretical foundations of these policies had been laid down by the Keynesian thought and Scandinavian programming theories. After all, planned economies of a Soviet type seemed to offer virtuous examples, thus pushing towards a more marked state orientation of the economic system.

After mid-Sixties, these orientations began to change, due to difficulty of controlling inflation. At the end of the Seventies, there was a radical change in the attitudes of two key countries, the United Kingdom and the United States, which found political expression in the election of Margaret Thatcher as Prime Minister, in the first country, and Ronal Reagan as President, in the second one. In practical terms, in these and other countries public intervention has drastically been reduced in both the allocative and distributive field. In some cases, intervention has taken different forms: e.g., attenuating tax progressivity, larger use of instruments imitating the market (such as assignment of trade permits for pollution control) or that, in any case, are based on its operation (such as money transfers) or, finally, reduce direct public involvement in the economy (by privatization and regulation of privatized firms).

To the root of these changes are also the liberist orientations that can be traced back to both the changes occurred in the working of the economy in the Sixties and some influential contributions in the economic doctrine, which, in some cases were inspired by those changes or took a cue from them. The main change in the economic attitudes that took place in the Sixties concerned the behaviour of economic agents, essentially as a consequence of a situation of full employment in all developed countries. The Phillips curve augmented by expectations devised by Milton Friedman can be thought of as an expression of such behavioural changes.

Robert Lucas' contribution, instead, can be traced back to an evolution of Friedman's position aimed at substituting the rational expectations hypothesis—which implies according to Lucas that the private sector can always neutralize public intervention, by making it ineffective—to the adaptive expectations assumption.

6 The Recent Changes of Direction and the Dangers for the World Economic Equilibrium

6.1 The Growth of China and the Instability of a Multilateral Equilibrium

The strong growth of China in the last decades at rates of the order of 10% and more has slowed down in the most recent years, but only with respect to its former pace, since it is still much higher than that of the other countries. This has led China to the first places in the world in terms of total GDP and also of foreign exchange reserves, with ambitious plans of further progress. In this country the number of poor, in absolute terms (i.e., in terms of the number of people earning less than 1.90$ per day at the purchasing power parity), has reduced from 835 millions in 1981 to 0.7 millions in 2015, with an incidence on population that has reduced over the same period from 84.0% to 5.11%. However, also higher incomes have increased and therefore inequalities have not lowered, but risen. In some cases, also the wages paid in China are higher than those in some European countries, such as those of the East Europe.

In technological terms, China is now a *leader* in many sectors. In not a few cases, primacy has been obtained due to skillful imitation of the technologies adopted in more advanced countries, which partly derives from imposition of specific rules to be obeyed by foreign direct investors in the country. For instance, full ownership of shares of foreign subsidiaries by foreign investors is prohibited in some sectors and access to the Chinese market is still exchanged with technology transfer, in spite of prohibition deriving from the membership of the WTO. On the other hand, China has encouraged foreign transnationals to create R&D centres in China, which implies the creation of positive externalities for the country, in particular by training of highly-skilled staff.

The increased relevance of Chinese economy and the recent financial crisis have accentuated requests of reform of the Fund's architecture. Following them, the G20 has decided in November 2010 to raise the Fund's quotas reserved to emerging countries. Since November 2015 the yuan is one of the reserve currencies composing the Special Drawing Rights (SDRs), a notional currency that has been created for valuing the funds lent to countries, making international payments or making interventions in foreign exchange markets. The countries accepting these Rights, which are usually those with a surplus in their balance of payments, increase the availability of resources and can make use of the SDRs for possible future deficits, in any case earning an interest. The yuan is in the basket of the SDRs together with the US dollar, the euro, the yen and the pound sterling, with a 10% weight. The decision of the IMF will have no other practical effect than making the renminbi to weigh for determining the value of the notional currency, which is given by those of the currencies composing it.

The decision to admit the renminbi in the determination of the value of SDRs is a signal important from a political point of view, as it stops the 'exorbitant privilege' of the United States once denounced by General De Gaulle. Chinese people think similarly to De Gaulle: some countries, in the process of losing their economic primacy, hold on to the privilege of their currency. From an economic point of view, this recognition of the importance of China's currency implies that this country will ensure stability of the value of the yuan, avoiding recourse to devaluation as a way to compensate for the decline of growth.

Apart from this, in a longer perspective the fear can arise that the multipolar equilibrium is unstable, as it was for similar situations in the past. Differently from the past, however, there could not exist the cultural and political factors that allowed a(n almost) smooth transition from a multipolar equilibrium (having the pound sterling and the dollar at the centre of the system) to an equilibrium with only one dominant currency (the dollar).

6.2 *Favourable and Unfavourable Conditions for the Chinese Growth*

Progresses made in the protection of intellectual property rights by China are among the favourable conditions. In fact, many patents have been registered in the last years and now China holds 20% of international patents, following immediately the United States, which hold 23% of them. But what is more impressing is that China has had a two-figures rate of growth of registrations for several years.

On the other hand, one has to recognise that China still silently violates some international agreements, causing European actions brought to the WTO and American retaliations, which, in turn have been followed by Chinese counter-retaliations. In fact, for technology transfer from the EU China imposes specific rules on industrial property rights and other intellectual property rights, which are different with respect to those that it applies to technology transfers between Chinese firms. This has two outcomes: first, to discriminate foreign owners of property rights; second, to limit the ability of foreign operators to protect some intellectual property rights in China, thus violating WTO rules. Officially as a retaliation against the Chinese violation of international agreements—but really as a part of neo-protectionist policies—during 2018 the United States have introduced a 25% tariff on 818 industrial and technologic Chinese products, ranging from cars to electronic products, medical equipment and part components of aircrafts—for a value of 34 billion dollars of imports. Beijing's reply has been to trigger similar sanctions on 545 American products, ranging from agricultural products to vehicles. In these conducts one can almost see the outbreak of a true trade war. In defense of China one can say that a country which is still emerging from a state of backwardness has more founded justifications than a country that is a leader in the world economy in adopting measures restrictive of foreign firms' penetration. However, acceptance by China of international rules that should envisage also similar situations implies compliance to them.

6.3 *The USA as Free Riders*

Also the Trump Administration is implementing a policy that threatens respect of international and regional agreements and heavily undermines the existing equilibrium.

In terms of the threats for the current international equilibrium, to the USA actions against China we have just mentioned one has to add the USA threat to Europe to raise tariffs against Europe and the threat of termination of the regional agreements stipulated by this country that has led to a new agreement of the NAFTA rules.

With respect to the WTO the Trump administrations seeks to justify what is really a violation of its rules, i.e., setting discriminatory tariffs on steel and aluminium, based on the argument that these are goods of relevance for national security, even if they are applied to imports from friendly countries such as Canada and Japan. To be true, Trump's decision could backfire against the USA (and other countries), since these two products are part of value chains from which also goods produced in the USA are obtained, which will lead to the rise in the prices of those goods.

Another action which is weakening WTO is Trump Administration's attempt to block the activity of the Dispute Settlement Body of this Organisation, by non-renewal of its members.

7 What Does the Future Hold?

7.1 Problems of Global Equilibrium in the Absence of a Global Government: Rodrik's Trilemma

A Turkish economist, Dani Rodrik, has highlighted mutual inconsistency of deep international integration, national sovereignty and democracy. This is the trilemma taking its name from the economist, which is illustrated in Fig. 2.

According to this trilemma two of these conditions optionally can be satisfied, but the third cannot at the same time. For instance, globalization and nation state imply renouncing to democracy, since national sovereignty would interfere with the consequences of international integration, which imposes following the indications of markets.

Accepting the couple democracy-nation state prevents full international economic integration, since that couple would lead to outcomes unwelcome for markets, due to the fact that the state would direct the economy in a direction different from that of markets. It was so in the Bretton Woods world, where capital movements were limited in order to leave some degrees of freedom for nation states.

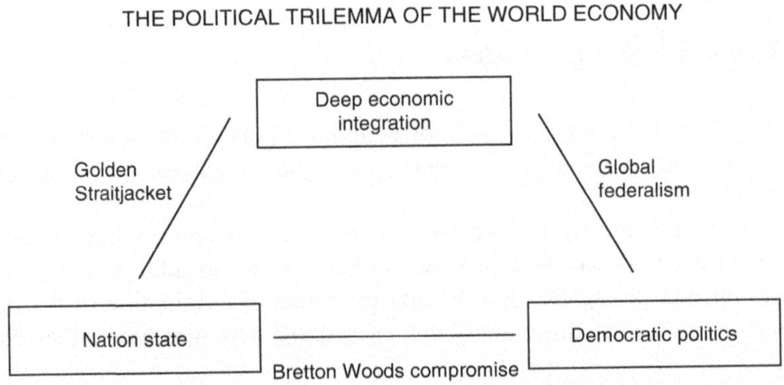

Fig. 2 The political trilemma of the world economy (Source: Rodrik 2002)

Finally, international integration and democracy at an international level would imply inexistence of the nation state, since the orientations of international organisations should be accepted.

7.2 *The Current Position and Perspectives*

As we have said, the institutions born at the end of WW II were inspired to a well-tempered liberism, but this principle has been disregarded, leaving instead room to an embittered liberism, which found its highest expression in the Washington Consensus. Then, the initial principle has largely been disregarded in practice.

The Bretton Woods institutions, as they have evolved, have led to positive results in terms of international integration and rates of growth of GDP, but are also responsible for financial and exchange crises, some of which are strictly tied to the request to countries in need of being financed to liberalise capital movements. To the World Bank one has also to ascribe the policy applied for some time to encourage growth of the country beneficiary of funds by prescribing lower taxes for the rich, with the idea that from this better life conditions would trickle down to the poor.

A number of critiques have been addressed to this embittered liberism, some of which have been accepted by international organisations.

If, until a few years ago the history of international economic relations after the war had offered a period indeed limited of tempered liberism followed by a longer period of almost complete liberalization, in the most recent years we have assisted to a new redirection of attitudes, with at least two countries, China and the United States, engaged in more or less open trade wars, in different ways.

On the other hand, the growth of China threats the current equilibrium, which is largely based on the USA hegemony, and a multipolar equilibrium is looming, where this country is joined by China. The two facts together, i.e., the trade war between China and the United States and the bi-polar equilibrium, can generate instability and dangers for the world economic equilibrium.

Therefore, the solution that many people would prefer in order to avoid the instability highlighted by Rodrik's trilemma, i.e. the association of international economic integration and global democracy—which also today shows deep faults, due to the government of almost all international organisations being firmly in the hands of the USA and, to some extent, of Europe—could receive further blows in the future and risks of being unrealizable. Further obstacles could derive from the diffusion of qualms towards transfer of power to supra-national levels and wide waves of populism in many advanced countries, especially Europe, and from widespread episodes of protectionism, in primis by USA and China.

Thus the future is not rosy and there are high probabilities of a return to the situations of trade wars that were typical of the interwar period.

From the well-tempered liberism which found its expression in the international organisations established at the end of WW II we have moved to an exacerbate liberism and it may be that the diffusion of populism and protectionism is the product, certainly, excessive, of that exacerbate liberism. Political commentators and historians attribute the election of Donald Trump and its economic policy to the economic crisis begun in 2007–8, which was born rightly from such exacerbate liberism. The populistic waves that afflict Europe, herald of similar closures, can have similar foundations.

If the world will stop in the path leading to commercial wars and closures of frontiers and will be able to reconstruct the season that led to the tempered liberism, is something desirable, but at present difficult to forecast.

References

Milanovic, B. (2013). Global income inequality in numbers: In history and now. *Global Policy, 4* (2), 198–205.

Rodrik, D. (2002). *Feasible globalizations*, NBER W. P. 9129, September.

United Nations. (2017). *Trends in international migrant stock: The 2017 revision*. New York: Documentation.

World Trade Organisation. (2009). *World trade report 2008*. Geneva.

GDP-Linked Bonds: A Proposal Worth Looking Into

Riccardo Barbieri Hermitte

Abstract GDP-linked bonds (GLBs) have been the subject of recurrent debates among economists and financial market participants but no major debt office has so far decided to launch them. Issuers remain skeptical about the practical feasibility or desirability of these securities, citing potentially high risk premia and various technical issues, including GDP data integrity and revisions, moral hazard and adverse selection. Against this backdrop, we discuss the key features of these securities and we argue that technical difficulties can be resolved, especially if standardized terms are agreed at the international level and statistical offices agree to publish dedicated GDP series. From the standpoint of investors, GLBs would provide direct exposure to GDP growth of a country or basket of countries without running into the more complex valuation issues that characterize equities. Long-dated GLBs also have the interesting feature that their duration may offset the impact of changes in GDP growth expectations. For issuers, regular issuance of long-dated GLBs would gradually raise the share of liabilities indexed to GDP. In the event of a major economic downturn, the debt-to-GDP ratio would rise more moderately than if all debt consisted of conventional bonds. In the limit case of GLBs accounting for the whole stock of public debt and without a price floor, the debt-to-GDP ratio would not change at all apart from the cyclical change in the budget balance.

Keywords Growth · Securities · Bonds · Debt · Country · Investors

---

The views expressed in this note are personal and do represent the official position of the Italian Treasury.

Paper presented at the Villa Mondragone International Economic Seminar, 25 June 2018, forthcoming in: Luigi Paganetto (editor), [title], Springer, 2020.

R. Barbieri Hermitte (✉)
Treasury Department, Ministry of Economy and Finance, Rome, Italy
e-mail: riccardo.barbieri@mef.gov.it

L. Paganetto (ed.), *Capitalism, Global Change and Sustainable Development*, Springer Proceedings in Business and Economics,
https://doi.org/10.1007/978-3-030-46143-0_5

1 Introduction

A recent survey of sovereign debt managers revealed a good deal of skepticism about GDP-linked bonds (GLBs), with respect to both technical aspects and potential investor interest. Issuers worry that weak demand for these securities would entail hefty risk premia, thus making them far too expensive.[1] In addition, the menu of government securities currently offered by Euro area countries is already quite broad, comprising not only bills and fixed-rate bonds, but also domestic and Euro-area inflation linkers and, in some cases, floating rate notes, not to mention various types of bills and bonds targeted to retail investors. Debt managers worry about the consequences of broadening the menu of government securities on market liquidity at a time of reduced gross issuance (and, for some time to come, central bank buying). Other oft-mentioned objections to GLBs, and more broadly state-contingent sovereign debt instruments, include practical complications relating to data integrity and revisions, moral hazard and adverse selection.[2]

However, there is no denying that, from the standpoint of sovereign issuers, GLBs would have superior defensive characteristics in the event of sharp recessions compared to conventional fixed-rate securities, as their redemption price would adjust to GDP developments. In a theoretical case in which all of a country's public debt was indexed to GDP, the public debt-to-GDP ratio would be fully hedged against unforeseen changes in nominal GDP growth, including sharp recessions such as the one that was caused by the 2008 global financial crisis.

GLBs have been the subject of renewed interest in academic and policy circles[3] of late. Most contributions concentrate on the tradeoff between the resilience of GDP-linked public debt to economic shocks and the risk premia that are likely to be demanded by investors, and hence the additional cost of funding likely to be borne by issuers. There is also broad recognition that, except for cases in which GLBs were introduced in the context of debt restructuring, it would take years of consistent GLB issuance, if not decades, to significantly improve the resilience of public debt to adverse economic shocks.

Meanwhile, work on more practical aspects of GLBs has been carried out by the Bank of England and a group of investors and has led to numerous versions of the so-called London Term Sheet (LTS).[4]

In this short essay, I shall focus on the aspects that are likely to determine the practical feasibility of GLBs, arguing that while these securities are no quick fix for high-debt countries, they are a debt instrument that should be seriously considered by issuers and investors.

[1]Key findings are summarized in OECD (2018).

[2]See the Executive Board Assessment in IMF (2017).

[3]See for instance Benford et al. (2016), Blanchard et al. (2016), Carnot and Pamies Sumner (2017), Kopf (2017), Manna (2017). Earlier contributions include Borensztein and Mauro (2004).

[4]ICMA (2017).

2 Technical Features of GLBs

In line with the LTS, I assume that a GDP-linker is a security whose principal and coupon payments are indexed to the growth rate of nominal GDP compared to a base year. At time t_0 the issuer sets a base for the semi-annual (or annual) coupon. The coupon paid at time t is equal to the ratio of the nominal GDP index at time t and at time t_0 (issuance), multiplied by the base coupon. The redemption price is equal to the notional times the ratio of nominal GDP at time t_N (maturity) and t_0. At any given point in time over the life of the GLB, a nominal GDP growth breakeven can be computed from the market price of the security, P_t, similarly to what happens in the case of inflation linkers.

3 Potential Development of Related Markets

It is fair to assume that if issuance of GLBs was large and sustained over time, that would lead to the development of trading in GDP swaps, as has been the case for inflation linkers. The availability of GDP swaps would create interesting cross-country trading opportunities. Investors may not only wish to position in GDP breakevens, but also trade them versus equity indices.

On the supply side, if spreads were attractive, corporate issuers too may become interested in issuing GDP-linked bonds. Nothing would prevent an issuer from a given country to issue a Euro-area GDP linker or an instrument linked to the GDP of another country if the cost was attractive once swapped into an interest rate exposure. Corporates may also wish to have liabilities linked to the GDP of a given country or basket of countries in view of the composition by country of their global turnover.

4 International Standards and Issuers' Coordination

GLBs could potentially be a useful instrument for debt management and a sovereign risk-reduction tool, especially if a relatively large number of Euro area or, better still, advanced economies adopted them. Isolated initiatives by individual EU member states would probably meet with limited success. A Pan-European initiative would seem more likely to succeed, as it would lead to larger issuance and secondary market trading volumes and provide investors with an opportunity to achieve a high degree of cross-country portfolio diversification. Although correlations between GDP growth rates of EU member states (and advanced economies) are high, the possibility of diversifying idiosyncratic risks would boost the appeal of GLBs and thus ultimately reduce issuer-specific risk premia.

Standardized features for GDP-linkers should also be agreed among sovereign issuers in order to minimize liquidity and novelty premia, hence the importance of agreeing on a specific term sheet from the outset.

5 A Long-Term Instrument and Issuance Program

GDP-linked bonds should be long-dated, e.g. 10 years and above. This is because of the likely investor base, the equity-like characteristics of the instrument and their technical features, notably the lagged indexation that may have to be adopted given the delay with which final GDP data are released by national statistical offices and Eurostat. Investors would thus take a view about long-term trends. Revisions to preliminary official estimates of nominal GDP would be more likely to balance out in the long run—an important consideration given that for the sake of timeliness coupon payments would be based on preliminary GDP estimates.[5]

A long maturity would also entail a higher financial duration of GLBs. As is the case for inflation linkers, high duration would offset or mitigate the risk of sharp falls in market prices of GLBs driven by changing GDP growth prospects. On the other hand, the price of a short-dated GLB would be more responsive to a sudden worsening in the economic prospects of a given country. With a higher duration, the drop in interest rates would offset the impact of changing GDP expectations on the price of the bond.

Issuance of GLBs should be large enough to ensure that, say, over a period of 10 years their share of outstanding debt would have a meaningful impact on the overall debt-to-GDP ratio and on default risk.

6 Which Countries Would Benefit from GLB Issuance?

As far as pricing is concerned, issuing GDP-linked bonds would be highly attractive for countries whose growth prospects are currently viewed favorably by economic forecasters and investors, and less attractive for the perceived growth laggards.

Indeed, if growth surprised on the upside issuers would end up paying higher ex post interest on GDP linkers compared with fixed-rate bonds; conversely, they would save on interest payments if nominal GDP growth disappointed original market expectations. Likewise, if nominal GDP outperformed expectations, the

[5]In EU countries, annual GDP data typically become 'final' after 2 years, frequently leading to significant revisions to the quarterly series. If one assumes that the final annual data are the 'true' level of GDP, the quarterly estimates that are available say, 4 months after the end of a quarter, should still be viewed as preliminary, which can lead to significant errors. It is thus important to balance out these errors over time (if one assumes that the estimation method does not suffer from systematic errors).

redemption value of the bond would rise compared to the par value, whereas it would fall below it if growth turned out to be lower than expected at issuance (unless the par value was guaranteed).

The defensive characteristics of GDP-linked bonds should thus be appealing not only for highly indebted countries, but also for issuers who currently enjoy low debt-to-GDP ratios. It is only the importance attached to hedging government debt against a sharp downturn in economic activity that may differ between the two types of countries.

7 'Sick Men' May Become Success Stories

An issuance program of GBLs should thus have a long-term horizon, regardless of temporary swings in growth sentiment. Countries' economic fortunes, and perceptions and expectations thereof, may change over time. At the time of writing, for instance, Germany is viewed as a highly successful country, but in the early 2000s it was dubbed the 'sick man of Europe'.

From Italy's point of view, pricing is the key concern at the current juncture given that the consensus view on Italy's long-term productivity and nominal GDP growth is rather pessimistic. On the other hand, the likes of Germany may not see a need to introduce a new debt instrument given the solid growth, sound public finances and ultra-low bond yields they have enjoyed in recent years. However, Germany would benefit from the positive assessment of its growth prospects among investors, economists and international organizations.

The European Commission's assessment of Germany's long-term real growth prospects (a 1.2% average growth rate for the 2020–2040 period) is lower than its estimate of potential growth in 2018–2019 (1.9%).[6] Even so, it would still yield a nominal GDP growth projection of at least 3%.[7]

The nominal GDP growth breakeven at which Germany would issue a hypothetical 20-year GLB would also depend on the risk premium demanded by investors. If the bond was issued at a breakeven of 3%, Germany's federal government would only be worse-off compared to issuing a fixed-rate Bund if nominal growth in the following 20 years exceeded 3%. (Germany's average nominal GDP growth rate since the inception of the euro, 1999–2017, has been 2.6%.)

---

[6]For long-term forecast see European Commission (2018); for the 2018–2019 forecast European Commission (2017).

[7]The introduction of GLBs would provide a strong stimulus to research on long-term GDP growth forecasting. The most common approach is to use a production-function method to estimate long-term real GDP growth and add to it an inflation (or GDP deflator) forecast broadly consistent with the central bank's inflation target. The average growth rate of Germany's GDP deflator in the 1999–2017 was 2.1%. A prudential forecast of Germany's nominal GDP growth in the 2020–2040 period would thus be 3.2% (1.2 real plus 2.0 deflator growth).

Hence, given the long-term nature of issuance programs, high-debt countries should be willing to incur an initially higher risk premium for the sake of improving their resilience, while low-debt, faster growing countries should take advantage of the relatively high GDP breakeven at which they would issue GLBs. The *ex post* cost of funding via GLBs would only exceed that of conventional fixed-rate bonds if nominal GDP growth exceeded current expectations.

8 A Source of Moral Hazard or a Stimulus to Reforms?

Critics argue that GLBs would create moral hazard because they would reduce the risk of explosive trends in a country's public debt-to-GDP ratio in the event of a confidence crisis and an ensuing recession. If governments did not fear the consequences of lax fiscal policies—the argument goes—they would break fiscal rules and run very large deficits. This argument overlooks the fact that in order to reach a high share of a country's total liabilities, debt offices would have to issue GLBs on an ongoing basis. If a given country ran unsustainable fiscal policies, markets would penalize it via a combination of higher risk premia and lower long-term GDP breakevens. Moreover, debt offices would continue to issue conventional fixed-rate bonds alongside GLBs, thus governments would not be able to run ultra-loose fiscal policies without incurring higher sovereign spreads.

In addition, given that the key variable for GLB pricing is *expectations* about long-term nominal GDP growth, governments that rely on GLB financing would have a strong incentive to implement credible growth-enhancing structural reforms. Governments may temporarily get away with far-fetched economic policy programs, but they would not be able to mislead consistently the markets over the course of a multi-year issuance program.

9 The Way Forward

Preliminary work has already been carried out in order to assess potential investor interest and the absorption capacity of global markets with respect to sizable issuance programs. This work should be broadened, involving all the main categories of investors, not only pension funds and mutual fund managers, but also sovereign wealth funds, official reserve managers and hedge funds.

A number of technical issues must also be resolved, including:

- Standardization of securities and indexation method (Term Sheet);
- Standardization of statistics for GDP-linked bonds in coordination with Eurostat;
- Minimum issuance as a share of gross funding programs.

The significant differences in terms of debt-to-GDP ratios and borrowing requirements currently existing among Euro area countries would complicate the coordination of a large-scale issuance program, not least in terms of issuance volumes. In addition, countries that issue larger amounts in absolute terms and relative to their annual issuance programs would probably end up paying higher premia. However, a concerted program would also engender active trading and arbitrage in ‘GDP breakevens’, which would help compress such premia.

As far as the quality and reliability of GDP statistics are concerned, investors would probably trust the independence of European statistical offices, supervised by Eurostat, and of those from other advanced economies.

The issue of GDP data revisions is complicated: there is a trade-off between timeliness of the GDP index used to calculate coupons and its accuracy, as quarterly GDP data are subject to significant revisions over time. If feedback from issuers and investors suggested that coupons should be semi-annual and that indexation lag should be relatively short, one would have to use quarterly instead of annual GDP data. In order to find a compromise between timeliness and accuracy, the third revision of the data, which is normally released between 85 and 120 days after the end of a given quarter, could be used to compute the relevant GDP index. For the purpose of GDP-linked bonds, those data would remain the relevant GDP figures. The statistical office would thus have to maintain and publish a dedicated nominal GDP series excluding subsequent revisions.

10 Conclusions

GLBs would improve the resilience of the public finances in the event of unforeseen economic downturns. As such, they would improve public debt sustainability. The insurance against economic downturns provided by GLBs would be particularly relevant for Euro area countries given the fiscal setup of monetary union (ECB independence and transnational status, as well as the no-bailout clause).

GLBs should preferably be long-term securities, for reasons that range from neutrality with respect to GDP revisions to the desirability of high duration.

Although OECD issuers have so far given the GLB proposal a cold reception, further preparatory work on commonly agreed standards for the securities and the statistics involved is thus warranted. A concerted EU effort, preferably in cooperation with other advanced economies, would raise the chances of successful introduction of GLBs. The interest that the European Commission has taken in the subject is thus welcome.

While Say’s Law (supply creates its own demand) may well apply to this case, gauging potential investor is the key priority at this stage. The oft-heard argument that GLBs would already exist if investors were genuinely interested in these securities does not hold water: only if a large and regular issuer started offering the new instrument, or a group of countries decided to move, would the GLB market and accompanying derivatives develop, thus compressing risk, liquidity and novelty premia. Gauging investors’ interest is clearly the priority. Indeed, while academic economists will keep probing the question of risk premia on GLBs, for market practitioners the proof of the pudding remains in the eating!

Finally, it is important to realize that only in the undesirable scenario of debt restructuring would a country be able to reach quickly a high share of GLBs over total outstanding debt. The attainment of greater debt resilience would otherwise take years of consistent and sizable GLB issuance.

References

Benford, J., Best, T., & Joy, M. (2016). *Sovereign GDP-linked bonds, financial stability paper no. 39*, Bank of England, September.

Blanchard, O., Mauro, P., & Acalin, J. (2016). *The case for growth-indexed bonds in advanced economies today*, Peterson Institute for International Economics, policy brief 16–2, February.

Borensztein, E., & Mauro, P. (2004). The case for GDP-indexed bonds. *Economic Policy, 19*(38), 165–216.

Carnot, N., & Pamies Sumner, S. (2017). *GDP-linked bonds: Some simulations on EU countries.* European Commission Discussion Paper 073, December 2017.

European Commission. (2017). *European economic forecast, Autumn 2017.* Institutional Paper 063, November 2017.

European Commission. (2018). *The 2018 ageing report: Underlying assumptions and projection methodologies*, European Economy 2018.

ICMA. (2017). *Introduction to the London term sheet for a GDP-linked bond*, 14 March 2017.

IMF. (2017). *State-contingent debt instruments for sovereigns*, IMF policy paper, May 2017.

Kopf, C. (2017). *The case for GDP-linked securities, presentation at BMCG meeting*, 7 February 2017.

Manna, M. (2017). *A formal introduction to GDP-indexed bonds*, mimeo, 13 October 2017.

OECD. (2018). *Sovereign borrowing outlook for OECD countries.*

Strengthening Disaster Resilience: A Microdata Perspective

Gero Carletto and Raka Banerjee

Abstract Each year, 206 million people on average are affected by disasters. Between 2000 and 2018, 84% of those 206 million people lived in Asia, home to 55% of 60,000 disaster fatalities worldwide. Unfortunately, most of the studies on disaster resilience available today are small-scale and *ad-hoc*, preventing generalization to the national level or broader contexts. Furthermore, the value of these types of analyses is often constrained by various data limitations. Within this paper, we briefly explore four areas of survey methods that offer the potential to improve the utility of microdata on disaster resilience: survey content, level of data collection, longitudinal data, and geospatial data integration. While not an exhaustive list of current data gaps, improving data availability and quality in these respects could have far-reaching impacts on the knowledge base to increase global disaster resilience for the world's most marginalized and vulnerable populations.

Keywords Data · Disaster · Household surveys · Measurement · Microdata · Resilience

JEL Codes H84 · Q54 · C81 · C83

This paper summarizes comments presented on the report "Asian Development Outlook 2019: Strengthening Disaster Resilience" at the XXXI Villa Mondragone International Economic Seminar (June 25–27, 2019).

G. Carletto (✉) · R. Banerjee
World Bank Group, Washington, DC, USA
e-mail: gcarletto@worldbank.org; rbanerjee3@worldbank.org

L. Paganetto (ed.), *Capitalism, Global Change and Sustainable Development*, Springer Proceedings in Business and Economics,
https://doi.org/10.1007/978-3-030-46143-0_6

1 Introduction

Each year, 206 million people on average are affected by disasters. Between 2000 and 2018, 84% of those 206 million people lived in Asia, home to 55% of 60,000 disaster fatalities worldwide (ADB 2019b). Unsurprisingly, while disasters pay no heed to income status, low-income countries tend to suffer more severe consequences in terms of fatalities as well as economic impacts relative to high-income countries. Within countries, the effects are similar: the poorest and most vulnerable segments of a population are those most likely to be severely impacted by disaster. Thus, it is vitally important to improve our understanding of the distributional impact of disasters at both the international and the intra-national level. Unfortunately, vulnerable populations often tend to be the most difficult to reach in data collection efforts, as they often live in marginal areas which are less accessible.

Several studies have been carried out using household survey data to understand the micro-level impacts of disasters in low-income contexts. A recent household survey on flooding in Indian cities found that in the absence of social protection or insurance, poor families were disproportionately affected by the disaster relative to rich households and were thrust further into indebtedness and poverty (Patankar 2020, forthcoming). These difficulties are compounded by evidence showing that poorer households are less able to take advantage of coping mechanisms for shocks, such as insurance and/or credit (Castells-Quintana et al. 2018).

The evidence provided by such studies is critical for governments and policymakers to identify needed policy interventions that can improve disaster resilience for their populations. Unfortunately, most of the studies on disaster resilience available today are small-scale and *ad-hoc*, preventing generalization to the national level or broader contexts. Furthermore, the utility of these types of micro-analyses is often constrained by data limitations. Within this paper, we explore four key areas in which improvements in microdata are needed: survey content, the level of data collection, longitudinal data, and geospatial data integration. While not an exhaustive list of current data gaps, improving data availability and quality in these respects could have far-reaching effects on the knowledge base to increase global disaster resilience for the world's most marginalized and vulnerable populations.

2 Survey Content

As with many areas of microdata, improving the measurement of disaster resilience requires refining definitions, increasing standardization, and validating better methods for the measurement of various key inputs that make up disaster resilience, as well as the outcomes and impacts of disasters.

2.1 *Defining and Measuring Resilience*

In conversations about disaster risk, disaster recovery, and disaster resilience, definitions can vary considerably, leading inevitably to variations in what is measured. Much has been written about resilience in recent years, but the international community has not yet come to a consensus on defining the term, perhaps because of the need to situate the understanding of resilience within specific contexts. In previous writings, we have used the definition articulated in the first publication of the Resilience Measurement Technical Working Group of the Food Security Information Network (FSIN) (Carletto et al. 2015). The working group defined resilience as the "capacity that ensures stressors and shocks do not have long lasting adverse development consequences" (Constas et al. 2014).

However, a more commonly used definition today comes out of the Sendai Framework for Disaster Risk Reduction, signed in 2015 and endorsed by all Asian Development Bank members in the Ulaanbaatar Declaration of 2019. The United Nations (UN) Intergovernmental Expert Working Group on Indicators and Terminology, in the course of its work to clarify the Sendai Framework's priorities, provided a working definition of resilience as "the ability of a system, community, or society exposed to hazards to resist, absorb, accommodate, adapt to, transform, and recover from the effects of a hazard in a timely and efficient manner, including through the preservation and restoration of its essential basic structures and functions through risk management." Operationalizing this into a workable measure is no easy task, requiring an expansion of both the breadth and the depth of data collected to measure resilience.

Increasing resilience is considered to be just one of four goals in the larger paradigm of Build Back Better (BBB), which came into prominence after the 2004 Indian Ocean tsunami. The other BBB goals, as defined by the abovementioned UN working group, are revitalization of livelihoods, revitalization of economies, and revitalization of the environment. To operationalize these goals, the working group proposed four distinct aims: safety, speed, inclusiveness, and opportunity.

- *Building back safer* requires recovery efforts to prioritize reconstructing infrastructure and environments in ways that enhance public safety in the event of another disaster, using both "hard" engineering approaches as well as "soft" policies such as establishing mechanisms to strengthen social ties.
- *Building back faster* requires that recovery efforts be accomplished as quickly as possible (ideally without jeopardizing the other goals of the BBB framework).
- *Building back inclusively* requires recovery efforts to take fairness and inclusivity into account by consulting with local communities and key stakeholders to empower the most disadvantaged segments of society.
- *Building back opportunity* requires recovery efforts to focus on supporting improved social wellbeing and better economic opportunities as key inputs to ensuring a sustainable and long-term recovery process.

Of course, within these definitions, discrepancies continue to exist between different actors. For example, a recent World Bank report opted for "building back stronger" rather than "building back safer", choosing to underline the importance of ensuring "that destroyed assets are reconstructed to more resilient standards" (Hallegatte et al. 2018).

Household surveys can play a key role in capturing many of these aspects of increasing resilience. While safety of infrastructure is beyond the scope of a household survey, issues of inclusivity could be captured at the household and/or community level by incorporating questions asking whether anyone in the household is aware of consultations with the local community on recovery efforts, for example. Similar questions could also be asked regarding household members' knowledge of whether recovery efforts have incorporated components to improve social wellbeing and economic opportunities in the community. Even the aim of "building back faster" could be captured through a household survey that included a panel component or, as an alternative, retrospective data. Thus, assuming thorough consultations are held with stakeholders to define the precise aspects of resilience to be measured and proper instruments are designed, household surveys are not only viable but well-suited to capturing disaster resilience in a wide variety of settings.

2.2 *Increasing Standardization*

The variation in definitions of resilience is mirrored in the variation seen in shocks modules used in different surveys. Table 1 from Heltberg et al. (2015) highlights the need for the international community to adopt and support a standardized set of guidelines for collecting microdata on shocks.

In addition to the above variation, shocks modules also differ widely in terms of reference periods, question wording, itemization of shocks experienced, inclusion of multiple levels of shocks, inter alia (Carletto et al. 2015). Without the necessary standardization on these aspects, meaningful cross-country comparisons are rendered impossible.

2.3 *Improving Methodology*

To strengthen the measurement of disaster resilience, it is important to reliably assess the impact of disasters, both direct and indirect. Direct impacts include those that affect wealth (assets; capital), resources (crops; the environment), and health (death; injury; disease). Indirect impacts include losses tied to foregone income or revenues, loss of alternative livelihood options, and so on.

Unfortunately, in many cases, the aftermath of a disaster poses significant roadblocks to accurate measurement. For example, measures of wealth in the form of productive assets usually include one's dwelling, which is often measured based on

Table 1 Typology of questions asked in shock and coping modules

	Experienced shock	Stock timing	Multiple shocks	Costs (type of loss)	Costs (currency)	Did others experience it	Severity	Individual who was affected	Has the household recovered	Coping type	Coping ranking
Afghanistan	x						x		x	x	
Bangladesh	x										
China	x			x	x				x	x	
Iraq	x									x	
Lao PDR	x	x		x	x				x	x	x
Malawi	x			x	x		x			x	x
Malives	x	x	x	x	x					x	
Mexico	x	x	x					x		x	
Nigeria	x	x	x				x	x		x	
Peru	x			x	x				x	x	
Sudun	x			x			x			x	
Tajtikitan										x	
Tanzania	x	x		x	x	x	x				
Uganda	x	x		x	x					x	
Uzbekistan	x						x		x	x	x
Vietnam	x	x			x			x	x	x	

Source: Heltberg et al. (2015)

projected rents in the housing market. However, in the case of places most prone to disasters, such markets are very thin, if not entirely missing, complicating measurement prospects. Measuring the indirect impact of foregone income can also be challenging, especially when the most vulnerable segments of a population tend to work primarily in the informal sector, where measurement is more difficult even in times of relative calm.

Another measurement challenge concerns a common source of livelihood in the developing world: subsistence agriculture. Strides have been made in bringing new sources of data to bear on long-time measurement challenges. For example, global nightlight maps have provided an innovative way to capture economic development at a scale beyond the scope of traditional household surveys (Weil et al. 2012). Unfortunately, such maps cannot shed light on the large-scale economic activity taking place on smallholder farms, nor on the potential economic impact when those farms are devastated by disaster. Thus, there remains an urgent need to devise better measures to capture disaster-related losses in agriculture with speed, precision, and at the requisite spatial granularity.

Finally, focusing exclusively on the areas directly affected by disasters is likely to underestimate their full impact on the broader community. Neighboring or adjacent areas that lie outside of the "eye" of the calamity are nonetheless likely to have been affected, for instance, through changes in labor markets or prices. Therefore, obtaining a more complete picture of disaster impact requires a broader scope for data collection efforts than may initially seem necessary.

3 Level of Data Collection

Just as disasters affect people at all levels of society, data collection on disaster impact and resilience must also capture data at multiple levels: the community, the household, and the individual.

3.1 Community-Level Data

The lack of readily available community-level data on disaster resilience represents a missed opportunity. Collecting this data should be a relatively low-hanging fruit; household surveys such as Living Standards Measurement Study (LSMS) surveys regularly include community modules to collect information on common resources, infrastructure, access to services, and social systems. Given the far reach of natural disasters, collecting community-level data serves as an opportunity to gain significant insight into the widespread impacts of disasters in a more efficient and cost-effective way. Currently, there are ongoing efforts to improve community-level instruments by increasing their ability to link to other data sources, such as administrative data and geospatial data, as well as non-traditional data sources such as mobile phones and sensors. However, to gain a better understanding of community constraints and opportunities in the face of natural disasters, much more work is needed to improve, harmonize, and better integrate field-based community-level data collection with other data sources.

3.2 *Household-Level Data*

Most of the methodological work done in the area of disaster resilience data to date has been at the household level. At present, the frequency, timing, and samples of household surveys in most low-income countries are often inadequate to capture the socioeconomic impacts among affected households when disasters strike. A recent study of the impact of floods in Malawi illustrates the enormous advantage of having a system of periodic household surveys in place to assess the effect of natural disasters (McCarthy et al. 2017). Nonetheless, the same challenges that apply to household surveys in developing country contexts also apply and are further exacerbated in disaster contexts, where the necessary societal and physical infrastructure needed to carry out high-quality household surveys may not be in place. This may also require resorting to different modes of data collection, such as through phones and mobile applications.

3.3 *Individual-Level Data*

Finally, measuring disaster impact becomes even more challenging when attempting to understand individual-level impacts. For instance, individual-level data is necessary to understand how disasters affect women, youth, and other vulnerable groups. The statistics presented in the 2019 Asian Development Outlook on strengthening disaster resilience do not provide any disaggregation (ADB 2019a). The 2030 Sustainable Development Agenda offers one opportunity for improving the collection of individual-level data; for one, work done by the UN Evidence and Data for Gender Equality (EDGE) project offers new methods for collecting gender-disaggregated data on topics such as asset ownership and entrepreneurship (Kilic and Moylan 2016). These guidelines, and other ongoing efforts, could be a starting point for exploring the collection of disaster resilience data at the individual level.

4 Longitudinal Data

Our understanding of resilience necessarily incorporates the capacity of individuals and societies to respond to and recover from shocks over time; thus, it is essential that data on disaster impacts must be able to capture the effects and/or transformation of populations over extended time periods. While multiple cross-sections conducted at successive points during disaster recovery can provide some insight on a societal level, longitudinal or panel data is needed to understand the experience of households and individuals in the aftermath of a disaster, as well as their trajectory to recovery. The wide availability of mobile phones today facilitates the implementation of follow-up survey efforts that can be more easily anchored to baseline data collection. In situations where panel data cannot be collected due to certain constraints, synthetic panels can be constructed; however, the still-experimental nature of this area of work makes it a less optimal choice.

Due to the sustained investment and capacity needed to conduct a longitudinal survey, collecting panel datasets often requires some form of institutional support and stronger capacity. For example, the Living Standards Measurement Study—Integrated Surveys on Agriculture (LSMS-ISA) initiative supported by the World Bank has worked with several countries in sub-Saharan Africa to conduct multiple rounds of panel household surveys. Other prominent panel survey examples are often supported by academic institutions, such as the collaboration between Yale University and the University of Ghana on a 15-year panel survey in Ghana, or survey work done by Michigan State University in Zambia. Despite the efforts of these institutions, much more longitudinal data is needed, particularly in low-income settings. Given the extended time period of such data collection efforts, it is crucial that the instruments are flexible enough to accommodate changes in response to evolving needs. It is also critical that the microdata are made widely and publicly available so as to maximize use.

If collecting panel data or conducting repeated cross-sections is not possible, the use of retrospective data based on recall may be a viable solution. Evidence shows that particularly for salient events—and the significant impact of a disaster certainly falls into that category—recall methods can produce unbiased estimates (Beegle et al. 2012).

5 Geospatial Data

The fast pace of growth in the area of satellite technology has led to a profusion of cheaply available high-resolution geospatial data from a wide range of sources. This eliminates any possible doubts around incorporating systematic georeferencing into household surveys and administrative data; it is a requisite, and can be done inexpensively with minimal time and skills. By georeferencing household surveys, one allows for the integration of vast amounts of geospatial data with household socioeconomic data, greatly expanding the potential and richness of a given survey dataset. This can be of particular relevance to the analysis of disasters, for which the availability of spatial data over an extended period of time is key.

Household surveys can also be used to ground-truth and calibrate measurements from satellite technology. For example, the combination of remote sensing with traditional data sources like household surveys has been demonstrated to improve the accuracy of crop yield measurement in Uganda as well as to improve estimates of economic wellbeing in Sri Lanka (Engstrom et al. 2017; Lobell et al. 2018).

Aside from its direct utility, geospatial data can also help improve other aspects of data collection. For example, geospatial data can facilitate the repurposing of broken administrative data systems (such as in the field of agriculture, where administrative data is widely understood to be of such low quality as to be unusable) to obtain estimates at a higher resolution with greater accuracy. Furthermore, geospatial technology can make it easier to incorporate a panel dimension into a survey: incorporating georeferenced data facilitates both tracking households over time as well as integrating multiple data collection efforts.

6 Conclusions

In this short paper, we have tried to highlight selected data issues which, if addressed, may improve our ability to monitor and understand the disaster resilience of individuals, households, and communities. Natural disasters affect a large number of lives and, if recent events reflect the changing environment, are likely to increase in strength and frequency in the future.

Data sources like household surveys, particularly when utilized in conjunction with other data sources, are critical tools for monitoring and understanding the socioeconomic impact of disasters and can help inform policies to strengthen the resilience of affected or at-risk populations. Disasters are bound to affect different segments of the population in different ways, but the poor and most vulnerable are likely to suffer the most. Measuring the differing distributional impacts of disasters and clarifying various pathways to recovery is crucial to minimizing the damaging effects of these adversities.

Natural disasters do not only uproot the lives of people in the immediately affected area. They can have long-lasting consequences on areas far away, through forced displacement and migration as well as other labor market impacts. Understanding this larger picture and properly quantifying the full impact when disasters occur will require more and better data. For household surveys to become more useful and relevant in the field of disaster resilience, we must collectively work towards identifying new ways in which these survey instruments can be repurposed, adapted to various contexts, and integrated with other data sources. Ultimately, we hope that addressing the issues identified in this paper will expand our global understanding of how the world can Build Back Better for its most marginalized and vulnerable populations.

References

Asian Development Bank. (2019a). *Asian development outlook 2019: Strengthening disaster resilience*. Manila: Asian Development Bank.

Asian Development Bank. (2019b). *Recent significant disasters in the Asia and the Pacific Region*. Infographic. Accessed February 4, 2019, from https://www.adb.org/news/infographics/recentsignificant-disasters-asia-and-pacific-region

Beegle, K., Calogero, C., & Himelein, K. (2012). Reliability of recall in agricultural data. *Journal of Development Economics, 98*(1), 34–41.

Carletto, C., Banerjee, R., & Zezza, A. (2015). *Household data sources for measuring and understanding resilience*. Resilience Measurement Technical Working Group. Technical Series No. 3. Rome: Food Security Information Network.

Castells-Quintana, D., Lopez-Uribe, M., & McDermott, T. K. J. (2018). Adaptation to climate change: A review through a development economics lens. *World Development, 104*, 183–196.

Constas, M., Frankenberger, T., & Hoddinott, J. (2014). *Resilience measurement principles: Toward an agenda for measurement design*. Resilience Measurement Technical Working Group. Technical Series No. 1. Rome: FSIN.

Engstrom, R., Hersh, J. S., & Newhouse, D. L. (2017). *Poverty from space: Using high-resolution satellite imagery for estimating economic well-being*. Policy Research Working Paper 828. Washington, DC: World Bank.

Hallegatte, S., Rentschler, J., & Walsh, B. (2018). *Building back better: Achieving resilience through stronger, faster, and more inclusive post-disaster reconstruction*. Washington, DC: World Bank.

Heltberg, R., Oviedo, A. M., & Talukdar, F. (2015). What do household surveys really tell us about risk, shocks, and risk management in the developing world? *Journal of Development Studies, 51* (3), 209–225.

Kilic, T., & Moylan, H. (2016). *Methodological experiment on measuring asset ownership from a gender perspective (MEXA): Technical report*. Washington, DC: World Bank.

Lobell, D. B., Azzari, G., Marshall, B., Gourlay, S., Jin, Z., Kilic, T., & Murray, S. (2018). *Eyes in the sky, boots on the ground: Assessing satellite- and ground-based approaches to crop yield measurement and analysis in Uganda*. Policy Research Working Paper 8374. Washington, DC: World Bank.

McCarthy, N., Kilic, T., de la Fuente, A., & Brubaker, J. (2017). *Shelter from the Storm: Household-level impacts of, and responses to, the 2015 floods in Malawi*. Policy Research Working Paper 8189. Washington, DC: World Bank.

Patankar, A. (2020, forthcoming). *Characterization of impacts of natural disasters on households and small businesses in India with specific reference to extreme precipitation events*. Asian Development Bank.

Weil, D. N., Henderson, V., & Storeygard, A. (2012). Measuring economic growth from outer space. *American Economic Review, 102*(2), 994–1028.

The European Globalisation Adjustment Fund: Easing the Pain from Trade?

Grégory Claeys and André Sapir

Abstract The European Union created the European Globalisation Adjustment Fund (EGF) in 2007 to assist workers negatively affected by globalisation in their search of a new job. The EGF was an acknowledgment that the EU, which has exclusive competence over trade policy, needed to assume some responsibility for the economic displacement due to globalisation. This article attempts to evaluate the EGF programme after 10 years of activity. Our evaluation addresses both its political visibility and its economic effectiveness. We find that the programme was visible in the sense that EGF beneficiaries tended to work in large firms and that their dismissals were reported in the media. The economic effectiveness of the programme is more difficult to evaluate because the available data is insufficient. Estimates, however, suggest that only a small proportion of EU workers who lost their job due to globalisation received EGF financing. Unfortunately, it is also impossible to assess whether workers who received EGF assistance did better in their job search than those who did not receive assistance. We make three recommendations to improve the programme: (1) collect more and better data to allow a proper evaluation of the programme; (2) revise the eligibility criteria to qualify for EGF assistance and the co-funding rate for the contribution by low-income regions; and (3) enlarge the scope of the programme beyond globalisation to also assist workers displaced by intra-EU trade and offshoring that result from the working of the single market, which is also an exclusive competence of the EU.

Keywords Globalisation · Economic crisis · Labour market · European · EU Budget

---

G. Claeys (✉)
Bruegel, Brussels, Belgium

Conservatoire National des Arts et Métiers (CNAM), Paris, France
e-mail: gregory.claeys@bruegel.org

A. Sapir
Bruegel, Brussels, Belgium

Université libre de Bruxelles (ULB), Brussels, Belgium
e-mail: andre.sapir@bruegel.org

L. Paganetto (ed.), *Capitalism, Global Change and Sustainable Development*, Springer Proceedings in Business and Economics,
https://doi.org/10.1007/978-3-030-46143-0_7

1 Introduction

Conventional wisdom in economics is that trade benefits countries through lower prices, greater product variety, better resource allocation and fuller exploitation of scale economies. All countries gain from trade, though some might gain more than others. Within countries, however, while most individuals stand to gain from trade, some might lose out, typically workers whose jobs are displaced by trade liberalisation.

It was the recognition that trade liberalisation, though good for the country, might hurt some workers, which led the United States Congress to establish the Trade Adjustment Assistance (TAA) programme under the 1962 Trade Act authorising the US to participate in the Kennedy Round (1964–67) of multilateral trade negotiations. Initially meant to provide income support to workers losing their jobs because of trade negotiations, the TAA programme was amended by the 1974 Trade Act to aid workers certified by their local state labour department as having been negatively affected by increased imports. The programme has remained in effect ever since. Although its economic effectiveness has been questioned, there is broad agreement that the TAA programme has played an important political role in obtaining the consent of Congress for trade liberalisation.[1]

For a long time, no such programme existed at European Union level for the simple reason that EU member states typically have much more generous welfare states than the US and are therefore better able to cope with the 'pains from trade'. What has long existed at EU level, however, is the European Social Fund (ESF), an instrument of the EU budget with a role assigned by the EU Treaty "*to render the employment of workers easier and to increase their geographical and occupational mobility within the Union, and to facilitate their adaptation to industrial changes and to changes in production systems, in particular through vocational training and retraining*."

However, unlike to the TAA programme, the ESF was not targeted specifically at workers affected by increased imports. Nor does it help equally workers across the entire territory of the Union. Instead it was designed to deal with industrial changes in general and to assist mainly workers in relatively low-income regions.

With the European Globalisation Adjustment Fund (EGF), established in 2006 and operational since January 2007, the EU now has an instrument broadly comparable to the TAA programme. The EGF provides financial assistance to facilitate the re-integration into employment of workers who have lost their jobs as a result of globalisation—defined as a substantial increase in imports into the EU, a serious shift in EU trade in goods or services, a rapid decline of the EU's market share in a given sector, or the offshoring of activities to non-EU countries—provided these redundancies have a significant adverse impact on the local, regional or national economy, regardless of whether they occur in high- or low-income regions.

[1] See, for example, Guth and Lee (2017) or D'Amico and Schochet (2012).

The creation of the EGF was a response to the rapid increase in globalisation and was a political acknowledgment that the EU, which has exclusive competence over trade policy, needs to assume some budgetary responsibility for the economic displacement that globalisation entails. Since the ESF only amounts to a tiny fraction of social expenditures by EU member states, it was recognised from the start that the EGF could assume only a relatively modest budgetary responsibility and that it needed, therefore, to be both politically visible and economically sensible.

The EGF programme originally ran for the entire duration of the 2007–13 Multiannual Financial Framework (MFF), the 7-year programming cycle of the EU budget. It was renewed in 2013 for the duration of the 2014–20 MFF. The financing for the EGF, currently capped at 150 million euros per year, comes from unused ESF money, of which it represents a tiny fraction. To give an order of magnitude, the ESF budget for 2014–20 is 84 billion euros, or 12 billion euros per year and close to 10% of the entire EU budget. The amount currently permitted for the EGF represents therefore a little over 1% of the ESF budget or 0.1% of the total EU budget.

In this chapter,[2] we evaluate the EGF programme after 10 years of activity and in the context of the negotiations on the 2021–27 MFF, which started in the Spring of 2018 with the publication of the European Commission's first detailed proposal. We describe the programme (Sect. 2), outline its functioning since its creation until 2016 (Sect. 3), evaluate its political visibility and economic effectiveness (Sect. 4) and make recommendations on how it can be improved (Sect. 5).

2 The EGF: Aims and Procedures

The EGF programme has changed substantially since its introduction in January 2007. The original legislation, dating from 2006, has been revised twice: in 2009, during the global financial crisis, and in 2013, during the negotiations for the 2014–20 MFF.

The EGF's original objective was to co-finance—together with national authorities in charge of implementing programmes at the local level—policies to facilitate the re-integration into employment of workers made redundant as a result of globalisation. The co-funding rate has changed several times since the creation of the Fund. The EGF share was 50% in 2007 and 2008, 65% during the crisis from 2009 to 2011, 50% again in 2012 and 2013 and 60% since 2014.

Given their contingent nature, there is no precise commitment for EGF expenditures in the Multiannual Financial Framework. The MFF only defines an annual cap for EGF expenditures from the EU budget, which are drawn from unspent ESF resources. This cap was initially set at 500 million euros for the 2007–13 budgetary

[2]This chapter was previously published as a working paper in the Bruegel Policy Contribution series as Claeys and Sapir (2018).

cycle, but was reduced to 150 euros million for 2014–20 because annual expenditures never reached more than this amount during the previous period (see Sect. 3).

In terms of procedures, to start an application, interested parties (regions, individuals, employers or representative organisations) should contact their EGF national coordinators. Given that the design and implementation of the programme is done at national level, potential applications are first screened by the member states. To be eligible for EGF financing, applications need to fulfil three basic requirements.

First, applicants must show that the redundancies are the result of globalisation. In 2009, the scope of the EGF programme was enlarged to redundancies resulting from "*the global financial and economic crisis*". When the EGF legislation was revised again in 2013, it was decided that during the 2014–20 budgetary period, the scope of the programme would continue to cover situations arising from the "*continuation of the global financial and economic crisis [. . .] or as a result of a new global financial and economic crisis*".

Second, individual applications must concern a fairly large minimum number of workers. The original number was 1000 redundancies in a particular (large) firm or in a group of smaller companies located in one or two contiguous regions. The threshold was reduced to 500 in 2009 and has been kept at that level since. Although both the original and the subsequent legislation allow for a lower minimum number in exceptional cases, the fairly large number of redundancies required to qualify for the programme suggests that political visibility, though not a stated objective, is an obvious goal of the EGF.

Third, EGF applications must be intended to finance active labour market measures to re-train and re-employ redundant workers. Such financing is obviously subject to a time limit. Originally, the implementation period for the measures covered by a successful application was 12 months. The period was extended in 2009 and has remained at 24 months since.

Applications must be submitted by interested parties to their national authorities, which must first verify that the eligibility criteria are met. Within 12 weeks of the announcement of the redundancies, national authorities must submit the application to the European Commission, which has then 2 weeks to respond to the national authorities, typically requesting additional information that the national authorities must provide within 6–8 weeks. The Commission then has 12 weeks to make a final determination of whether the application meets the required criteria. Once approved by the Commission, the application is submitted to the Council of the EU and the European Parliament, which must give their approval to commit EGF expenditures on each individual case as the EGF is formally outside the MFF process.

3 The EGF: Facts and Figures

Two data sources are available to describe the functioning of the EGF since its creation: the list of EGF applications made available by the European Commission's employment, social affairs and inclusion directorate-general (DG EMPL), and the biennial reports[3] presented by the European Commission to the European Parliament and the Council describing the activities and accomplishments of the EGF.

DG EMPL uploads regularly on its website the list of EGF applications, which provide case-level data on the sector of the company applying, the application date, intervention criteria, budgeted national and EGF contributions and the number of targeted workers. This is available for all applications (accepted, withdrawn, rejected) from 2007 to 2016.

The Commission's biennial reports include additional case-level information on completed programmes for which the Commission has received a final case report from national authorities (which they are required to file with the Commission 6 months after the completion of each EGF-financed programme[4]). The latest Commission report at time of writing was presented in October 2017. These reports give information on the characteristics of targeted workers (age, gender), the number of workers actually helped, actual amounts spent and number of workers re-integrated into the labour force at the end of the implementation period. The most recent cases examined in the latest report date from 2014. This means that for the more recent cases, the only information available is the information on budgeted values and targeted workers contained in the list of applications provided by DG EMPL.

Between 1 January 2007 and 31 December 2016, 147 applications to the EGF, covering 140,545 redundant workers, were approved. This represents an average of about 15 applications and 15,000 workers per year, i.e. 1000 workers per application. Only 16 of the 147 applications concerned situations with fewer than 500 redundancies.

The total number of redundant workers who actually benefitted from EGF financing was 91,505, but this figure only applies to the period 2007–14 (for which data on actual financing is available at the time of writing), during which 113,904 redundant workers were approved as eligible for EGF financing. The difference between eligible workers and actual beneficiaries reflects the fact that about 20% of the eligible workers had already found a new job by the time their EGF application was funded.

Although originally designed to deal with the consequences of globalisation, the EGF has actually been used more often to deal with redundancies caused by the

[3]This report switched from annual to biennial in 2014. See European Commission (2008, 2009, 2010, 2011, 2012, 2013, 2015, 2017).

[4]The publication of these reports, which could be useful to assess the EGF, is not mandatory and only a few member states publish them (see for instance Ireland: http://egf.ie/final-reports-for-completed-egf-programmes/), Thus, we do not have access to most of them.

Table 1 Distribution of EGF cases (2007–2016)

	Number	Percentage
Total EGF cases	147	
Number of SMEs cases	68	46%
of which Globalisation	*24*	35%
of which Crisis	*44*	65%
Number of single firm cases	79	54%
of which Globalisation	*46*	58%
of which Crisis	*33*	42%

Source: European Commission (2018)

economic and financial crisis: 52% of cases, covering 51% of the redundant workers and awarded 55% of the funding related to crisis fallout rather than globalisation. Table 1 shows how EGF cases are distributed both in terms of justification (globalisation or crisis) and in terms of type of firm concerned (redundancies in a single large firm or in a group of small and medium sized companies, SMEs).

Figure 1 gives the total number of targeted redundant workers and the total funding committed under the EGF, broken down between 'globalisation' and 'crisis' for each year from 2007 to 2016. Several points should be noted. First, the number of targeted workers made redundant because of globalisation has ranged between about 3000 and 13,000 per year and their EGF funding has been between 14 million euros and 52 million euros per year. Second, the number of workers made redundant as a consequence of the crisis has ranged between 0 and 24,000 per year and their EGF funding has been between zero and 115 million euros. Third, the maximum amount of funding awarded by the EGF in any year was 132 million euros in 2010, when 115 million euros was associated with the crisis and only 17 million euros with globalisation, which explains why the annual EGF envelope was lowered from 500 million euros to 150 million euros in 2013. Finally, the average amount of EGF funding awarded per worker over the period 2007–16 was 4219 euros. Given that over the period, the average share of co-financing provided by member states was 42%, it means that each redundant worker eligible under the EGF received on average 7274 euros in active labour market services.

Figure 2 shows how the total number of redundant workers and the total funding committed under the EGF from 2007–16 was distributed to EU countries. Ten countries (Belgium, Denmark, Finland, France, Germany, Greece, Ireland, Italy, Netherlands and Spain) accounted for respectively 83% of all the redundant workers and 87% of all the funding under the EGF programme. Second, among the main users, there were substantial differences in the reason for seeking help from the EGF: for some countries the main reason was globalisation (Belgium, Finland and Germany), for others it was the crisis (Greece, Ireland and Netherlands) and for the rest (Denmark, France, Italy and Spain), it was a combination of globalisation and the crisis. Third, central and eastern European countries benefitted little or not at all from the programme, presumably because they lost few jobs because of their relatively sustained growth over the period. Finally, the United Kingdom, though presumably as much affected by globalisation and the crisis as other countries in western Europe,

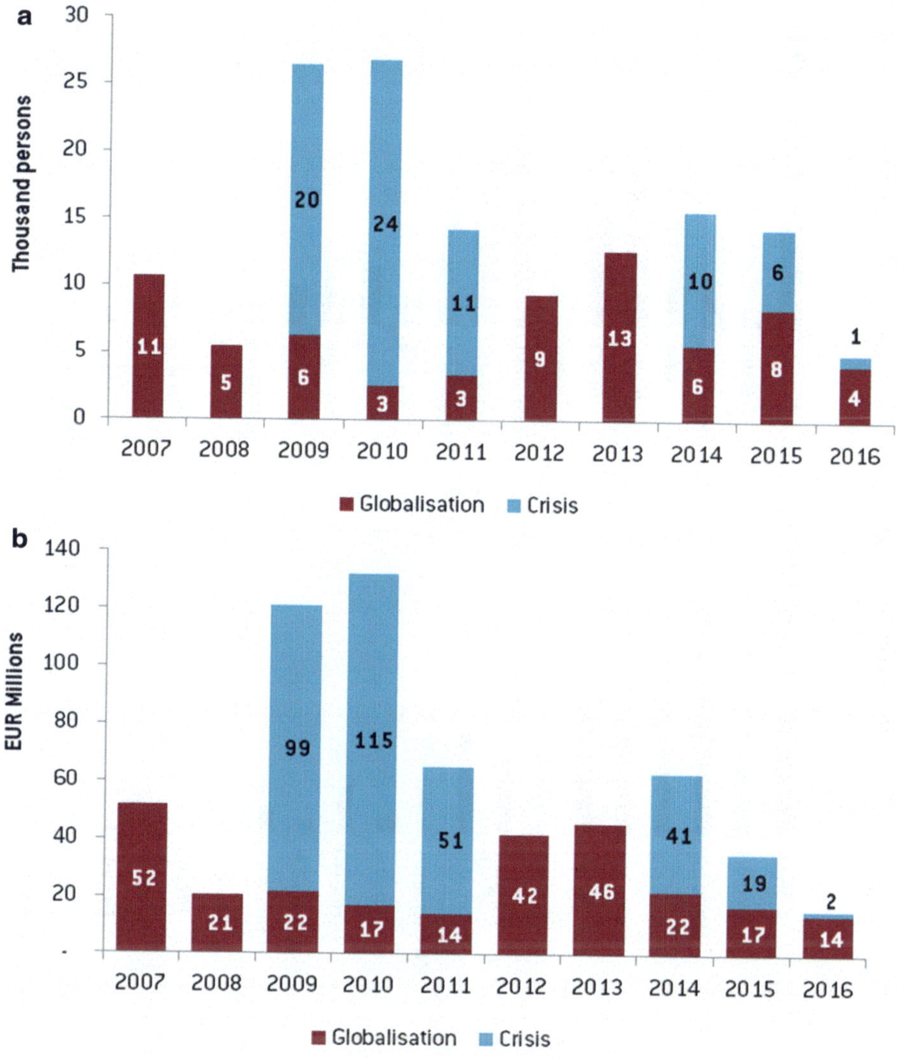

Fig. 1 Number of targeted workers and total amount committed by the EGF, by year and justification (2007–16). Source: European Commission (2018)

did not use the programme at all. This seems to have been the result of a deliberate decision by the national authorities.[5]

[5]The UK is the only country in the EU without a national coordinator for the EGF and the British government made it clear that it was not interested in participating to the EGF and that it did not want the EU programme to be renewed. See for instance in 2012: https://publications.parliament.uk/pa/cm201213/cmselect/cmeuleg/86-xvi/8605.htm. Moreover, we find in the ERM database (see

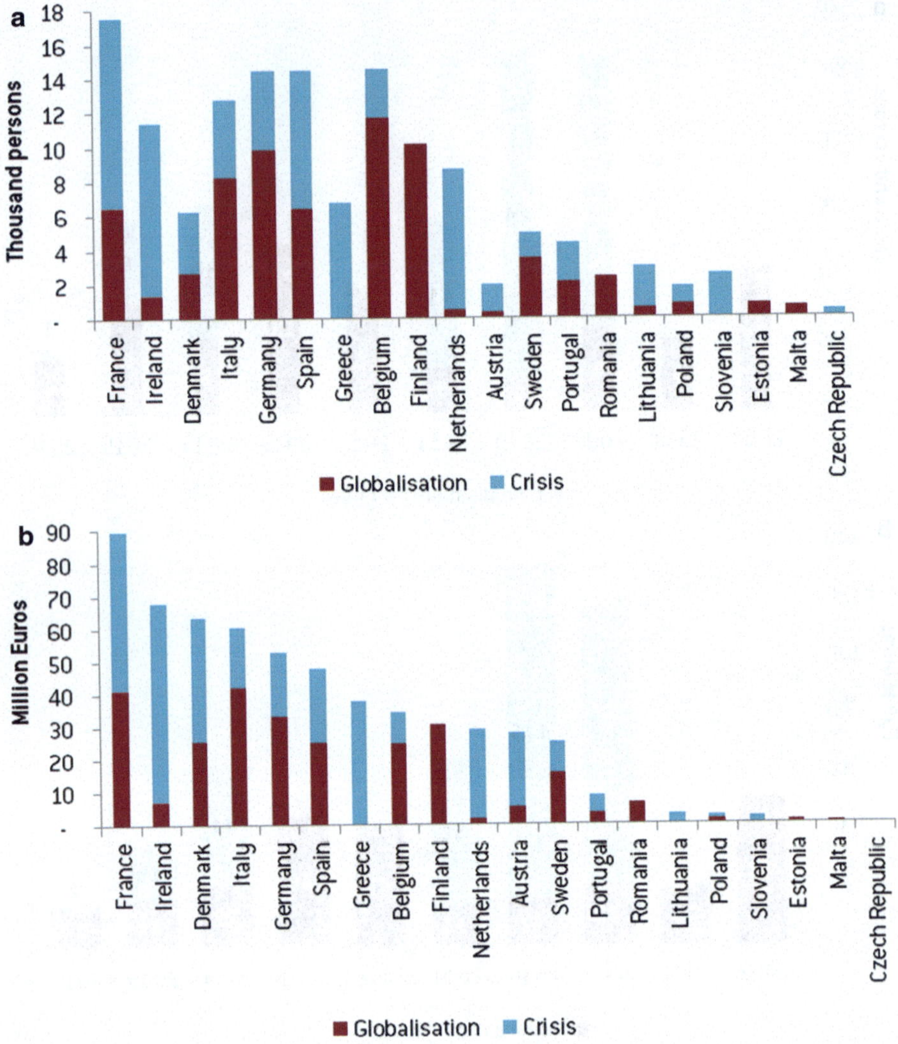

Fig. 2 Number of workers targeted and total amount committed by the ESF by country, 2007–16. Source: European Commission (2018)

next section) that at least 6 cases of extra-EU offshoring could have been eligible to EGF funding in the UK between 2007 and 2016.

4 The EGF: An Assessment

A proper evaluation of the programme requires answers to three main questions. First, if the EGF is above all a political tool, it needs to be visible. Has this been the case? Second, even if the EGF is mainly political, it also needs to be sensible from an economic perspective, raising the questions: what proportion of workers who became redundant in the EU during the period 2007–16 as a consequence of globalisation received helped from the EGF? And how well did the workers helped by the EGF do in regaining employment, compared to workers with similar characteristics who lost their jobs for similar reasons but did not receive EGF help?

From a political perspective, the visibility of EGF cases is essential. To evaluate their visibility, we use the European Restructuring Monitor (ERM) database, which provides data on large-scale restructuring events reported by the media since 2003. The database covers restructuring events affecting at least 100 jobs or 10% of the workforce at worksites with more than 250 employees. It should therefore contain information on all the single-firm EGF cases, since they involve at least 500 redundancies. For each restructuring event, the ERM database provides information on the name of the affected company, its size, location and sector, and the type of restructuring and number of jobs lost. The ERM database covers 15,465 events for the period 2007–16. All EGF cases, except for one single-firm EGF globalisation case, are included in ERM database, which suggests that single-firm EGF cases are, in most cases, highly visible.

In terms of whether the EGF programme has played a significant role in helping EU workers made redundant by globalisation, there is no existing data for the number of such workers. Lawrence (2014) suggests that 5% of job losses in the United States might be a consequence of globalisation. Assuming a similar proportion applies to the EU would imply that roughly 180,000 jobs are lost annually in the EU because of globalisation. How does that compare to the number of workers helped by the EGF? We found (Sect. 3) that on average for the period 2007–16, 14,054 workers qualified for EGF assistance, of which only 48% lost their jobs as a consequence of globalisation. Therefore, a little below 7000 workers made redundant by globalisation received EGF assistance each year, amounting to roughly 4% of the total of job losses ascribed to globalisation.

An alternative approach would to use the ERM database and to focus on redundancies reported in the media that involve at least 100 workers or 10% of the workforce at worksites with more than 250 employees. Unfortunately, it is generally not possible in this database to distinguish whether restructurings are the result of globalisation or other causes, unless restructurings involve offshoring, for which the ERM database distinguishes between intra- and extra-EU offshoring. The ERM database includes 30 cases for the period 2007–16 linked to extra-EU offshoring and involving at least 500 layoffs—meaning they were in principle eligible for EGF funding. However, EGF financing was received in only four of these cases. This suggests that only a small proportion of redundancies arising from globalisation—

even among large, politically visible firms that received media attention—received EGF support.

The question is therefore why the number of workers qualifying for EGF assistance is so small compared to the estimated number of workers losing their jobs because of globalisation, and why cases that could have been eligible for EGF assistance were not included. Beyond the fact that the amount of money available for the EGF programme is fairly small (which has not in itself been an issue given that the cap has never been reached since the creation of the EGF), a survey of national authorities in charge of the EGF programmes by Puccio (2017) indicates several potential reasons for its under-use and also its uneven deployment in different EU countries:

- Eligibility criteria might be too strict. First, the minimum threshold for the number of workers affected might be too high, especially for small countries or those where affected workers are mainly in SMEs. Second, the scope of the EGF might be too narrow: it excludes, for instance, redundancies generated by intra-EU trade and offshoring (while single market policies are also an EU exclusive competence).
- Administrative rules might be too complex and result in lengthy procedures, which might discourage applications.
- The level of co-funding by national authorities might be too high, especially compared to other EU funds. In particular, ESF-financed programmes provide between 50% and 85% (95% in exceptional cases) of total project costs depending on the relative wealth of the region.[6] Some central and eastern European countries might opt for the ESF—even for redundancies that might be generated by globalisation—over the EGF, which offers a lower co-funding rate.

Finally, we examine the effectiveness of the EGF in helping dismissed workers find new jobs. The available data indicates that the average re-employment rate of workers who received EGF assistance is 46% by the end of the implementation period (2 years since the revision of the EGF rules in 2009). However, it is important to note great variability, with re-employment rates ranging on average between 26% in Belgium and 92% in the Czech Republic (see Fig. 3).

The average re-employment rate of 46% means that, overall, only two-fifths of the workers eligible for EGF financing found a job within 2 years thanks to this financing. As discussed earlier, one-fifth had already found a job before the financing started and the remaining two-fifths had not yet found a job at the end of the 2-year EGF implementation period.

We cannot assess, however, if the 46% average re-employment rate can be considered satisfactory, because (1) we do not have data for individual workers who received EGF assistance (but only for the average worker in individual EGF cases); and (2) even if we did, we could not compare EGF-assisted workers with

[6]See ESF guidelines: http://ec.europa.eu/esf/main.jsp?catId=525

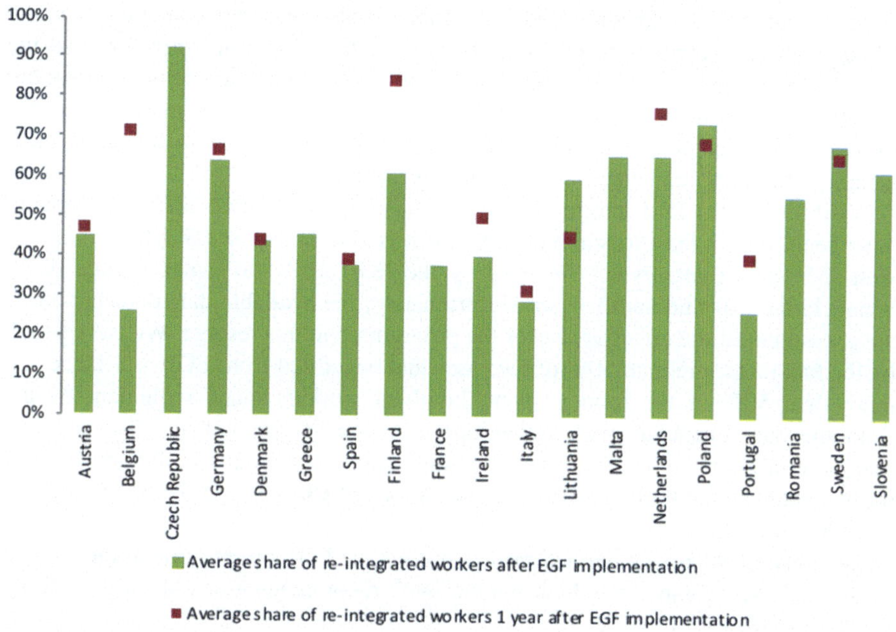

Fig. 3 EGF re-employment rate, 2007–14 (%). Source: European Commission reports to the European Parliament and the Council (2008 to 2016). Note: the EGF re-employment rate is calculated as the number of workers re-integrated into the labour market at the end of the EGF-financed programmes divided by the total number of beneficiaries. This data is available for 121 cases from 2007 to 2014. The average share of re-integrated workers 1 year after EGF implementation is based on 67 cases from 2007 to 2011, as reported by the ex-post evaluation of the EGF (Andrews et al. 2015).

equivalent dismissed workers available in data from Eurostat's European Labour Force Survey (ELFS) or EU Statistics on Income and Living Conditions (EU-SILC), because these two data sources provide information neither on the reason for dismissal nor on re-employment rates within a 2-year period.

5 Conclusions and Recommendations to Improve the EGF

Overall, the idea behind the EGF—to help correct the negative side-effects of globalisation with active labour market policies and to be seen to do so—was certainly valid in principle because globalisation creates losers as well as winners.

Politically, it was important for the EU to create a budgetary instrument that would enable it to assume some financial responsibility—even if relatively modest—for assisting workers displaced by trade liberalisation, considering EU trade policy is an exclusive EU competence. Amounts devoted to EGF programmes are necessarily very limited compared to those involved in member states' labour market policies, and in their education policies, which are also essential for coping with the labour market changes induced by globalisation. It is important therefore that the philosophy of the EGF programme, which is to give national authorities a small but targeted financial incentive to put in place or improve active labour market measures to assist displaced workers, rather than simply provide them with unemployment benefits, should be translated into effective measures.

Visibility of the EGF programme is an objective that matters and that seems to have been met, at least to some extent. But it is also essential that EGF money be well spent and that services financed by the EGF really do make it more likely beneficiaries will find another job. Unfortunately, the available data does not allow an assessment of the effectiveness of the programme in this respect. We were struck by the fact that more than 50% of the cases that benefitted from EGF assistance and more than 50% of the money spent, involved workers made redundant by the economic and financial crisis, rather than by globalisation per se. This suggests that there might have been significant administrative constraints in the operation of the programme for trade adjustment assistance, which were partly lifted to assist workers hit by the crisis.

We make three recommendations to improve the EGF programme in the context of the next MFF, some of which overlap with recommendations already made by Cernat and Mustilli (2017):

1. There is a need to improve the monitoring and evaluation of the programme by collecting more and better data. The present situation is clearly unsatisfactory because it does not allow a proper scientific evaluation of the EGF. The best approach would be to collect data at the individual level and not only at the case level.[7] This would allow evaluators of the EGF to use quasi-experimental methods by building a relevant control group that is similar to the treatment group (based on age, gender, education, experience, marital status, other unemployment services received, etc.). At the very least, final case reports should be made available and should be standardised in terms of measures undertaken and outcomes. This would have the additional advantage of making cases comparable so that member states can share good practices. It would also give some needed visibility to the EGF.
2. There is a need to revise the administration of the programme to increase its use. There are too many constraints on eligibility. First, one should envisage having no threshold (like the TAA) or at least a much lower threshold than currently. Obviously, this would imply that the amount of money needed for the programme could increase significantly,[8] but it would also mean greater equity in the

[7]Collecting data at the individual level in order to be able to undertake scientific *ex-post* evaluation was recommended by Wasmer and von Weizsäcker (2007) when the programme was launched.

[8]Coming back to our back-of-the-envelope calculation, if all 180,000 workers affected by globalisation were covered, this would represent a total budget for EGF-financed programmes of 800 million euro per year (assuming expenditure equal to the average amount spent in the period 2007–16, i.e. around 4200 €/beneficiary).

Table 2 Offshoring events reported in the ERM database and EGF cases included

	500 or more redundancies	Between 100 and 500 redundancies
Intra-EU	19 (3)	203 (5)
Extra-EU	30 (4)	152 (3)
Various locations	19 (3)	80 (0)
Total	68 (of which 10 EGF cases)	435 (of which 8 EGF cases)

Source: Bruegel based on Eurofound, European Restructuring Monitor (ERM) (available at: http://www.eurofound.europa.eu/observatories/emcc/erm/factsheets) and European Commission (2018)
Note: EGF cases included in parenthesis. There were eight EGF cases which, when matched with the ERM database, are labelled as 'intra-EU offshoring'. These eight cases were submitted to the EGF on the grounds of a serious shift in EU trade in goods or services rather than the offshoring of activities to third countries

allocation of funds between workers who work in large establishments, who tend to be eligible, and those working in relatively small companies who tend to be excluded, unless they are geographically concentrated. Second, the co-funding rate could be changed. The new rate should the same as the co-funding rate for ESF programmes (and thus be different for different regions, depending on their relative income levels) in order to avoid any disincentive to use the EGF, in particular in central and eastern European countries, where the national co-funding rate for the EGF is higher than for the ESF. Third, the European Commission could be more proactive in its management of the EGF. It could for instance use the ERM database to detect redundancy plans that meet the EGF eligibility criteria and suggest to national authorities that they could apply and put in place EGF programmes for these cases.

3. Finally, one should consider enlarging the scope of assistance for adjustment from globalisation to other policy-induced sources of adjustment, including intra-EU trade and offshoring, and the phasing out of activities in order to reduce carbon emissions (Tagliapietra 2017). The ERM database suggests that increasing the scope to intra-EU offshoring and reducing the threshold to 100 redundancies, could result in several hundred additional EGF cases (see Table 2).[9] The EGF could therefore become the EAF, the European Adjustment Fund, with expanded resources. Ideally these resources should be included in the MFF through the creation of a specific budget line (which would also have the advantage of speeding up the procedure). The additional resources for the European Adjustment Fund should come from the budget currently allocated to the European Social Fund.

Acknowledgement The authors would like to thank Raphaële Adjerad and Inês Gonçalves Raposo for their excellent research assistance.

[9]Reducing the threshold to 100 redundancies would also result in an additional 150 extra-EU cases.

References

Andrews, M., Weber, T., Pavlovaite, I., & Smith, R. (2015). *Ex-post evaluation of the European globalisation adjustment fund (EGF)*, ICF International.

Cernat, L., & Mustilli, F. (2017) *Trade and labour adjustment in Europe: What role for the European globalization adjustment fund.* Chief Economist Note. DG Trade: European Commission.

Claeys, G., & Sapir, A. (2018). *The European globalisation adjustment fund: Easing the pain from trade?* Bruegel Policy Contribution No 2018/05. Available at: https://www.bruegel.org/2018/03/the-european-globalisation-adjustment-fund-easing-the-pain-from-trade/

D'Amico, R., & Schochet, P. Z. (2012). *The evaluation of the trade adjustment assistance program: A synthesis of major findings.* Final report prepared as part of the evaluation of the trade adjustment assistance program for the U.S. Department of Labor, Employment and Training Administration, SPR Project No. 1213.

European Commission. (2008). *Solidarity in the face of change: The European globalisation adjustment fund (EGF) in 2007 – Review and prospects, communication from the commission to the European parliament and the council, COM (2008) 421 final.*

European Commission. (2009). *Report from the commission to the European parliament and the council on the activities of the European globalisation adjustment fund in 2008, COM (2009) 394 final.*

European Commission. (2010). *Report from the commission to the European parliament and the council on the activities of the European globalisation adjustment fund in 2009, COM (2010) 464 final.*

European Commission. (2011). *Report from the commission to the European parliament and the council on the activities of the European globalisation adjustment fund in 2010, COM (2011) 466 final.*

European Commission. (2012). *Report from the commission to the European parliament and the council on the activities of the European globalisation adjustment fund in 2011, COM (2012) 462 final.*

European Commission. (2013). *Report from the commission to the European parliament and the council on the activities of the European globalisation adjustment fund in 2012, COM (2013) 782 final.*

European Commission. (2015). *Report from the commission to the European parliament and the council on the activities of the European globalisation adjustment fund in 2013 and 2014, COM (2015) 355 final.*

European Commission. (2017). *Report from the commission to the European parliament and the council on the activities of the European globalisation adjustment fund in 2015 and 2016, COM (2017) 636 final.*

European Commission. (2018). *Summary of EGF applications – 2007 to date (Monthly update).* DG-Employment website, available at http://ec.europa.eu/social/main.jsp?catId=326, last visited on 26.1.2018.

Guth, K., & Lee, J. (2017). *Evaluations of the trade adjustment assistance program for workers: A literature review.* Executive Briefings on Trade, U.S. International Trade Commission.

Lawrence, R. Z. (2014). Adjustment challenges for US workers. In F. Bergsten, G. C. Hufbauer, & S. Milner (Eds.), *Bridging the pacific: Toward free trade and investment between China and the United States.* Washington, DC: Peterson Institute for International Economics.

Puccio, L. (2017). *Policy measures to respond to trade adjustment costs.* EPRS Briefing, European Parliament.

Tagliapietra, S. (2017) Beyond coal: facilitating the transition in Europe. *Bruegel Policy Brief*, Issue 5. Available at http://bruegel.org/2017/11/beyond-coal-facilitating-the-transition-in-europe/

Wasmer, E., & von Weizsäcker, J. (2007). A better globalisation fund. *Bruegel Policy Brief*, Issue 1. Available at http://bruegel.org/2007/01/a-better-globalisation-fund/

The Effect of Brexit on the UK Economy (So Far)

Sindri Engilbertsson and Gylfi Zoega

Abstract The political turmoil in the UK following the referendum on future membership of the European Union in 2016 provides a natural experiment for studying the effects of political uncertainty on the economy. We find that the subsequent confusion and infighting in British politics has not affected the real economy much—employment is at a historical high and output growth is positive—but there are some signs of slowing investment and house price increases. The stock market has also not been much affected although it did fall after the referendum of 2016. The main effect of the Brexit vote and the subsequent political developments is found in the currency market where news that make a hard Brexit more likely cause the currency to depreciate. We conclude that leaving the European Union without an agreement is likely to make the currency depreciate and the stock market fall while output declines. In contrast, leaving with an agreement that gives continued access to the Single Market would likely make the currency appreciate, the stock market rise and employment and output increase further.

Keywords Brexit · British economy

A preliminary version of this paper appeared as a working paper at the Birkbeck Centre of Applied Economics, BCAM 1907, 2019.

S. Engilbertsson
Department of Economics, University of Iceland, Reykjavik, Iceland
e-mail: sie49@hi.is

G. Zoega (✉)
Department of Economics, University of Iceland, Reykjavik, Iceland

Department of Economics, Mathematics and Statistics, Birkbeck College, University of London, London, UK
e-mail: gz@hi.is

L. Paganetto (ed.), *Capitalism, Global Change and Sustainable Development*, Springer Proceedings in Business and Economics,
https://doi.org/10.1007/978-3-030-46143-0_8

1 Introduction

The political turmoil in the United Kingdom beggars belief. Few would have believed that the British political establishment was capable of creating this level of chaos. Segments of both of its large political parties, especially the Conservative Party, spent decades stirring up antipathy towards the European Union without making any plans to leave the bloc. The government of David Cameron promised to hold a referendum before the 2015 elections to stave off a challenge from the emerging Independence Party and to please its own eurosceptics. The referendum was held in 2016—not because the government thought it was a good idea to leave the EU but because it had to deliver on a promise made for party-political reasons. In fact, such was its dislike for the Leave option that, in the months preceding the referendum, Cameron's government did not make any preparations for leaving the European Union.

A responsible government would have presented an alternative plan to leaving, so that voters would know what they were voting for and the government would know how to proceed if the majority of voters wanted to leave. However, this was not a responsible government. Its decision to hold a referendum without a plan of action has caused the subsequent political turmoil. In this paper, we do not delve into the political background of the referendum. Instead, we are interested in the effects of the turmoil on the economy. In essence, the entire episode is a natural experiment in the effect on economic activity of a political class, riven by personal ambitions playing with a country's interests and future.

An expanding literature in economics and political science focuses on the reasons for the outcome of the Brexit vote.[1] They include the effects of international trade on the distribution of income and regional disparities, the effect of immigration on unskilled service workers' income, and concerns about national identity and traditional values.[2] Crafts (2019) attributes the Brexit vote to the financial crisis of 2008 and the Conservative government's ensuing austerity policies after 2010, which mainly affected the regions in the north of England that voted to leave the European Union. Thus the leave vote can be interpreted as a protest vote against the austerity policies and, as such, a consequence of the financial crisis. Crafts goes so far as to suggest that if the UK had put some of the bankers in jail in 2009, the

[1]See Antonucci et al. (2017) who attribute the Brexit vote to the declining financial position of the middle class; Arnorsson and Zoega (2019) on the regional characteristics of the Brexit vote (older, less educated, lower income); Colantone and Stanig (2018) on global competition and Brexit; Fidrmuc et al. (2016) on the absence of a relationship between EU regional subsidies and the remain vote; Hobolt (2016) on the divisions in society; and Inglehart and Norris (2016, 2019) on the cultural factors behind the vote. Goodwin and Heath (2016) find that turnout was generally higher in more pro-Leave areas, which implies a greater intensity among those who wanted to leave than those voters who wanted to stay.

[2]Dustman and Frattini (2014) analyse the fiscal impact of immigration on the UK economy since 1995. They find that immigrants from the European Economic Areas (EEA) have made a positive fiscal contribution while non-EEA immigrants have made a negative contribution. Since 2000, the contributions have been positive throughout, in particular for immigrants from Europe.

wrath of the population might have been assuaged enough to stave off the Brexit vote.

One can place the Brexit vote into the context of a broader political movement that includes the election of the current US president in 2016 and the emergence of right-wing populist movements in Northern Europe and left-wing populist movements in Southern Europe. These movements share certain features, such as having charismatic leaders, challenging conventional wisdom, and defying international organisations. Right-wing populists tend to be nativist, while left-leaning populists tend to oppose capitalism and international finance rather than opposing immigration.[3] These movements have split traditional parties on the left and right into nationalistic groups, on the one hand, and those that champion trade and migration, on the other.[4]

But does this political turmoil matter for the economy? Perhaps the economy is immune to these developments. However, any change in the relationship between the UK and the European Union in the future is likely to have some economic impact, positive or negative, and expectations about this impact should affect today's asset prices, exchange rate, investment, and consumption.[5] The economy may not be immune, and the expected economic consequences may be dire. So how has the UK economy weathered the storm so far?

2 Political Uncertainty

If economic agents could see into the future, if they had perfect foresight, the post-Brexit world would affect present-day share prices, exchange rates, consumption, and investment. But clearly, they do not. If they could instead have rational expectations—taking into account the probabilities of a good outcome, a bad outcome, and everything in between—we could also use market prices to decipher what the future will probably look like. Share prices would reflect available information on the likelihood that the UK will leave the European Union and the likely effects of its staying or leaving on future growth, interest rates, capital flows, and so forth. However, these probabilities cannot be calculated because the UK has never left the European Union before; therefore, no available information can be used to assess the probabilities of exit and the many different arrangements that could conceivably follow EU membership. In other words, there is uncertainty about the future relationship with the European Union.

[3]See Rodrik (2018) on the difference between left-wing and right-wing populism.

[4]See Mudde (2004, 2017) and Müller (2015, 2016) on populism.

[5]Dhingra et al. (2016) analyse the economic consequences of the UK leaving the European Union. They find that while these will depend on which policies the UK adopts following Brexit, lower trade due to reduced integration with EU countries will likely cost the UK economy much more than is gained from lower contributions to the EU budget.

Frank Knight (1921) made the distinction between risk and uncertainty. Risk involves well defined probabilities that can be calculated from repeated instances that are identical, such as the repeated tossing of a coin. It can also involve statistical probabilities, such as the probability of death of an individual of a certain age. Uncertainty arises when the distribution of the outcome in a group of instances is not known because the situation in question is unique.[6] Political events such as Brexit involve uncertainty, not risk. For this reason, traders in the stock market and the currency market could be expected to find it difficult to assess the impact of the UK's leaving the European Union simply because it has never happened before and the final outcome is completely uncertain. The UK might end up belonging to the EU; it might enter the European Economic Area with Norway, Lichtenstein, and Iceland; or it might reach a bilateral trade agreement with the EU. It might also follow WTO rules.

In the following sections, we attempt to assess whether markets gauge the effect of Brexit to be positive or negative for the UK economy, and the strength of their response to news reports that make Brexit seem more or less likely. We start with the real economy.

3 The Real Economy

It may come as a surprise to many that the impact of the Brexit turmoil on the real economy, output, and unemployment has been muted so far. This is surprising because both consumption demand and investment demand depend on expectations of the future, and the profit outlook may have worsened due to the political turmoil.

For most of the 3-year period since the Brexit vote, the economy has been doing well. The unemployment rate is at its lowest since the early 1970s, and employment as a share of the working-age population is at a record high (see Fig. 11 in Appendix), while output growth has been respectable, although not very strong. The capital inflows that have generated persistent current account deficits since the mid-1990s have continued. Figures 1, 2 and 3 show that the British economy continues to exhibit largely the same features as before; there is steady growth in output, unemployment is low and declining, and there is a current account deficit of around 4% of GDP.

The economy is driven mostly by the services sector, such as banking; economic activity is concentrated in several large cities where house prices tend to be high; employment is booming; and there are continued capital inflows, owing most likely to the country's sound institutions, law and order, protection of property rights, and impartial courts.

However, there are indications that investment has fallen slightly in recent months, as is shown in Fig. 12 in the Appendix, probably due to Brexit uncertainty. This decline may indicate that further uncertainty about Brexit—not to mention a

[6]See Hill (1997) on political risk.

Fig. 1 Current account as a proportion of GDP

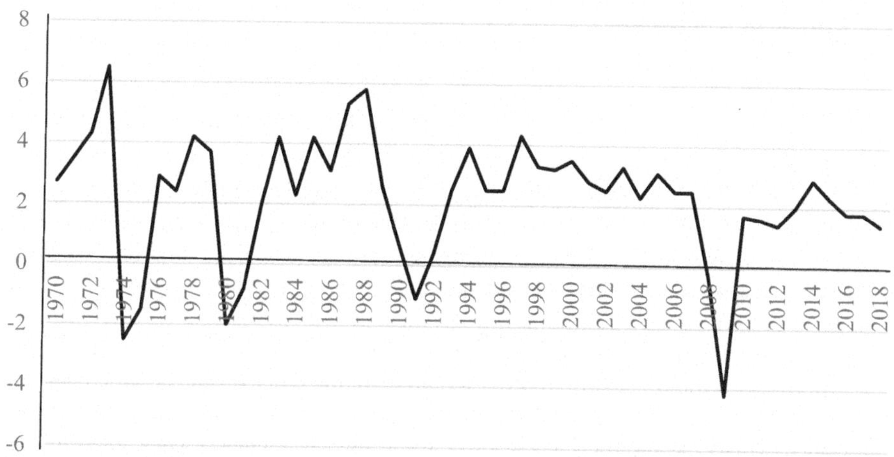

Fig. 2 Real GDP growth

no-deal exit—will cause output and employment to contract. In addition, house price inflation has eased, as is shown in Fig. 13 in the Appendix. However, it is noteworthy that an economy threatened by the immediate imposition of tariffs on inputs and output and the disruption of supply chains in manufacturing should manage to perform this well. Another consequence is the reduction in European immigration into the UK triggered by the rhetoric of the country's political class.[7]

[7]See Forte and Portes (2017)

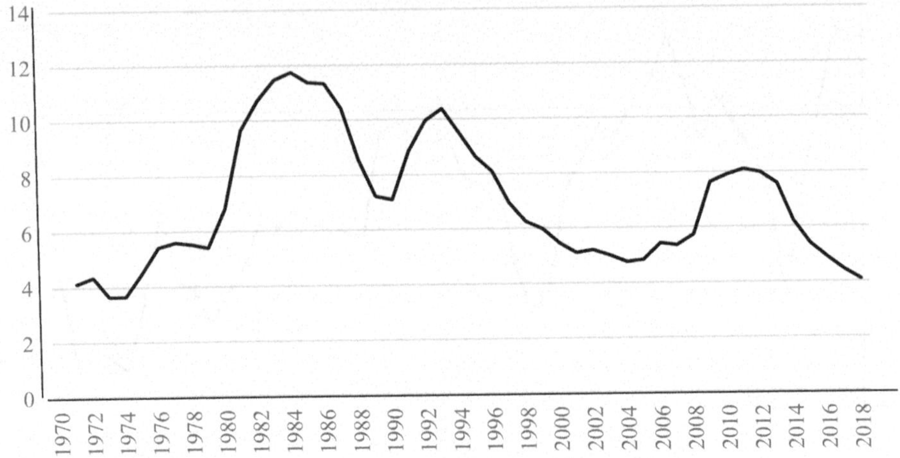

Fig. 3 Unemployment rate

4 The Stock Market

Even if the real economy is doing well, then, the stock market must surely have been hit by the Brexit uncertainty and the probability that tariffs will be imposed! We study the FTSE 250, which includes companies ranked 101–350 in terms of market capitalisation in the UK stock market, so as to omit the large international corporations included in the FTSE 100 that are less affected by Brexit (Wright 2016). The FTSE 250 is shown in Fig. 4 below, alongside the German DAX and the US S&P 500. The effect of the Brexit vote on the index seems to have been mild[8] when looking at data from the beginning of 2015 through 3 September 2019.

We took the logarithm of the stock market index for each day and calculated the first difference of the logarithm in order to calculate the daily proportional change in the index. Then we constructed indicator variables for changes greater than four, five, and six times the standard deviation of the series. The dates of large movements in stock prices were then compared to data on Brexit-related events in Parliament (House of Commons Library 2019) and news reports appearing in newspapers, both the day before and the day after the large change in the stock market. In this way, we attempt to determine which news reports were likely to lead to a rise in the stock market and which were likely to cause a fall.

Table 1 shows the days between 2 January 2015 and 3 September 2019 when the FTSE 250 changed by more than four standard deviations, the size of the proportional change, and the apparent cause of the change. There were 15 such large changes, 10 of them apparently Brexit-related. The fall in the index on 24 June 2016 was by far the largest change over this period. The remaining five large changes in

[8]FT250 data taken from investing.com

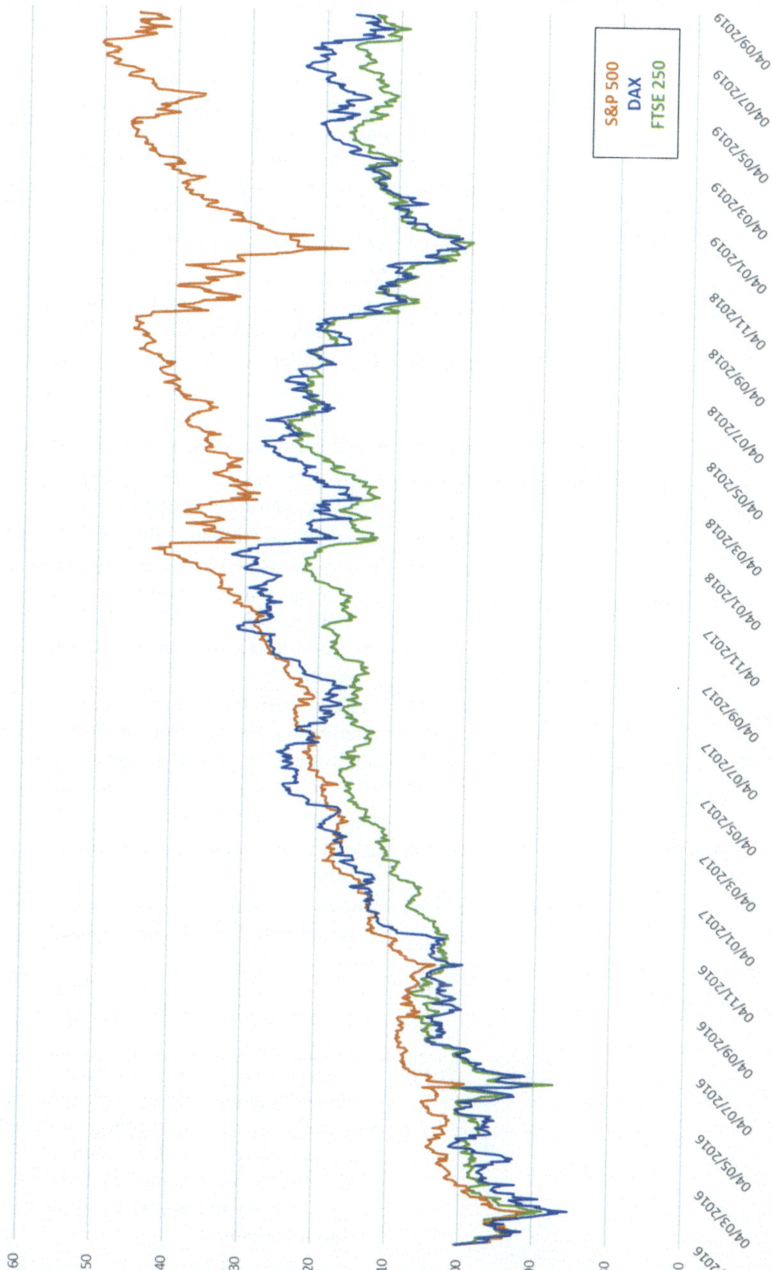

Fig. 4 FTSE 250 compared to the DAX and the S&P 500, 2 Jan 2016 through 3 Sep 19 (Source: Authors' calculations and investing.com)

Table 1 Large stock market movements and events between 2 Jan 2015 and 2 Sep 2019

Deviation exceeds $6*\sigma^2$	Deviation exceeds $5*\sigma^2$	Deviation exceeds $4*\sigma^2$	Change	Reason for deviation
		8.5.2015	0.0120	(7.5.2015) PM David Cameron is re-elected after promising a referendum on Brexit. He promises to fight to keep Britain in the EU and stakes his career on the outcome.
24.8.2015			−0.0174	Black Monday in China causes stock markets to tumble.
	25.8.2015		0.0134	Markets recover from China's Black Monday.
	20.1.2016		−0.0130	Market turbulence caused by falling oil prices along with pessimism in Davos.
	8.2.2016		−0.0139	News reports indicating economic slowdown trigger fears of an impending global recession.
		17.2.2016	0.0124	Markets recover from global recession fears.
		17.6.2016	0.0104	(16.6.2016) Labour MP, Jo Cox, is assassinated by a far-right extremist.
	20.6.2016		0.0140	Polls indicate growing "Remain" support.
24.6.2016			−0.0324	The results from the Brexit referendum are announced and were as follows: Remain: 16,141,241 (48.1%) Leave: 17,410,742 (51.9%). PM David Cameron announces his resignation.
27.6.2016			−0.0313	Markets open for the first time post-referendum. David Cameron discusses the implications of Brexit in Parliament.
	28.6.2016		0.0153	BoE responds to the vote with liquidity injections into the banking sector.
	29.6.2016		0.0138	Brexit panic recedes causing markets to jump.
		5.7.2016	−0.0104	Three large investment fund managers freeze their property funds in the UK due to uncertainty regarding Brexit.
	11.7.2016		0.0140	(11.7.2016) Confirmation that Theresa May will serve as the next PM. May did not support Brexit.
		6.12.2018	−0.0125	Escalation of the trade war between China and the US along with news of the arrest of Huawei's chief financial officer in Canada causes contraction in global markets. FTSE 250's drop is compounded by PM Theresa May's inability to pass her Brexit agreement through Parliament.

Source: Authors' calculations, investing.com, British newspapers, and House of Commons Library (2019)
Note: The standard deviation over the period was: $\sigma^2 = 0.00255$

the index all occurred before the Brexit referendum took place. The FTSE 250 spiked after the murder of Parliamentarian Jo Cox on 16 June 2016, rising by more than three standard deviations on the following day, most likely because the market thought that a leave vote was now less likely. On 20 June 2016, when the last opinion polls predicted that the Remain side would win, the index shot up by more than five standard deviations. After the results of the referendum became known on 24 June, and after markets reopened on 27 June, the index fell until the Bank of England decided to cut interest rates. The stock market responded positively to the appointment of Theresa May on 11 July but then remained stable until 6 December 2018, when the Huawei manager was arrested in Canada, intensifying the trade war between the US and China.

We also compared the FTSE 250 to the German index and the S&P 500 for the period 4 January 2016 through 3 September 2019.[9] Each index was given a base date of 4 January 2016. All three indices in Fig. 4 fall after the Brexit vote but recover quickly. Thereafter, they all increase until worries about the trade war and a possible global recession escalate at the end of 2018. The S&P rises more than the German and UK indices, the latter moving together. This masks an important development: UK companies have lost value in US dollar terms because of the depreciation of the pound sterling. Figure 5 shows the three indices measured in dollars. Here the FTSE 250 falls much more in comparison to the other two indices.

It has been shown that the effect of the possibility of the UK leaving the EU differs across companies and industries. Davies and Studnicka (2018) found that Brexit would have the strongest adverse effect on firms with supply chains in Europe and would affect smaller firms more than larger ones.[10] Moreover, Davies and Studnicka show that even though the stock market recovered after the initial drop in summer 2016, the businesses most strongly affected—that is, those relying on supply chains in Europe—did not recover their previous market value at the time of their writing.

5 Currency Market Reactions

The market most affected by the Brexit turmoil has been the currency market. The value of sterling falls when an exit from the European Union becomes more likely—not to mention an exit without a prior agreement—or when there is increased uncertainty about an exit.

Figure 6 displays the pound-euro and pound-dollar exchange rates for the period from 4 January 2016 through 3 September 2019. The change of 24 June 2016, when the results of the Brexit referendum were announced, is highlighted, as are the low points in the exchange rate. The pound-dollar exchange rate fell to its lowest post-

[9]Data taken from investing.com

[10]See also Corsetti and Müller (2016).

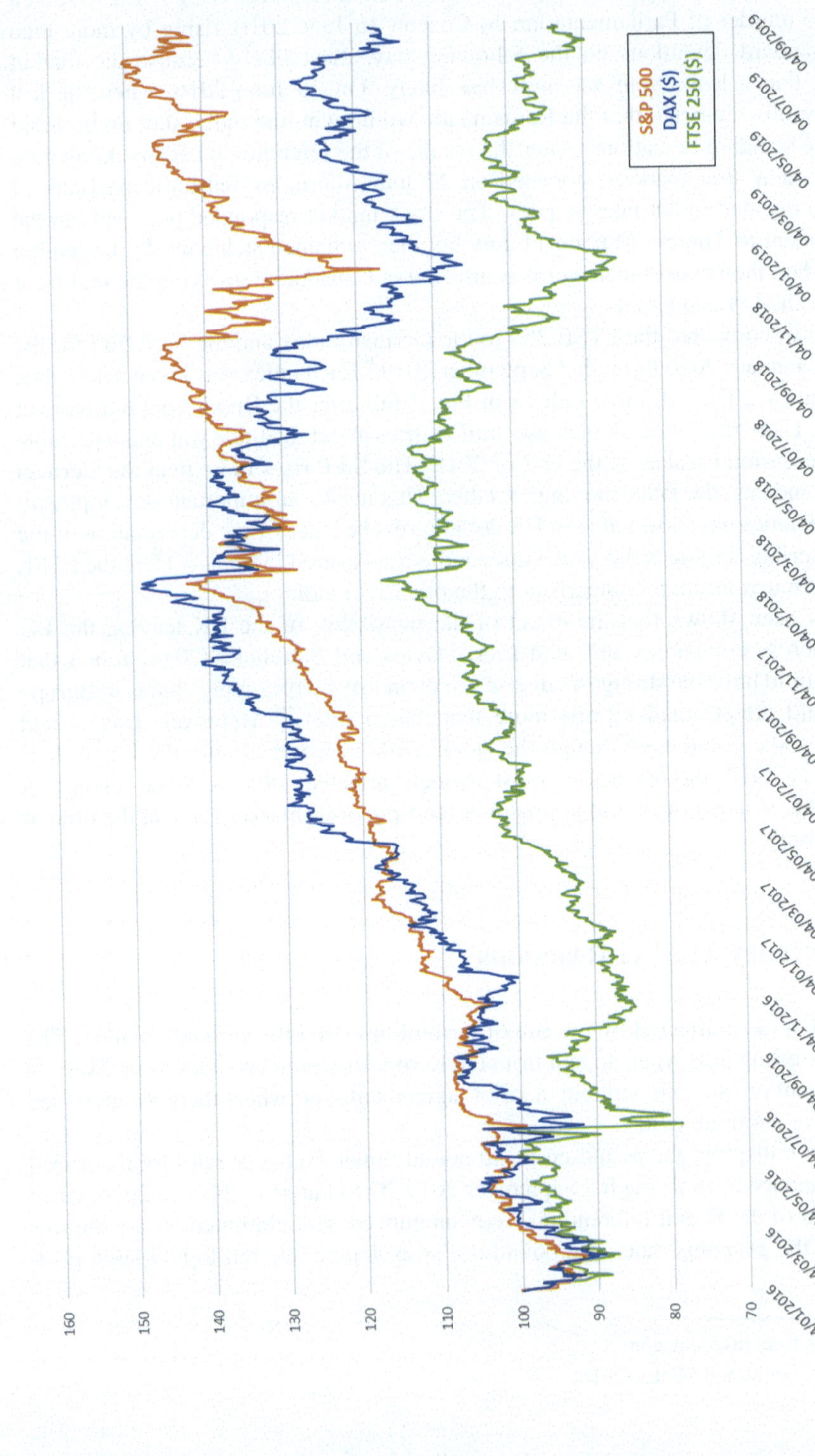

Fig. 5 FTSE 250, DAX, and S&P 500, in US dollars, 2 Jan 2016 through 3 Sep 2019 (Source: Author's calculations, investing.com and landsbanki.is)

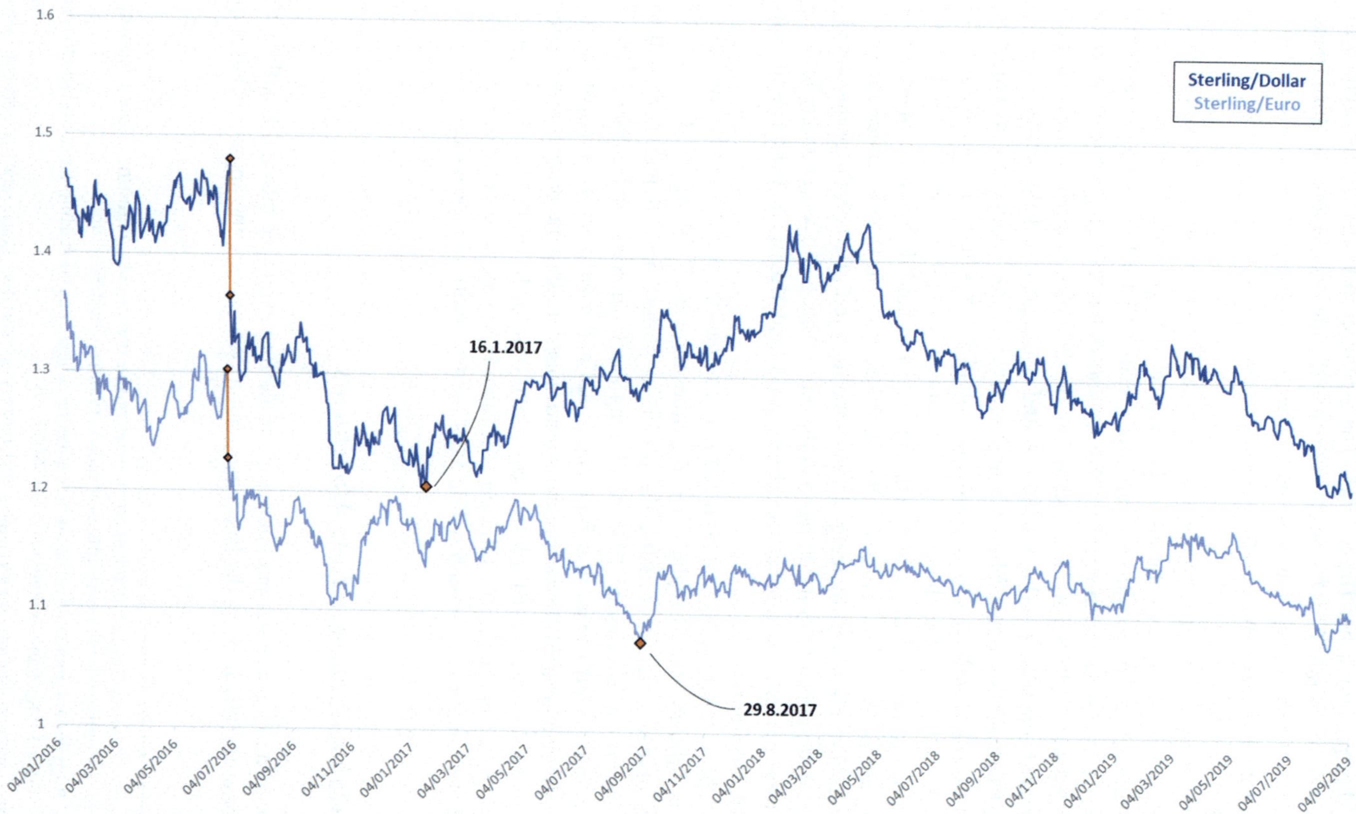

Fig. 6 Pound-euro and pound-dollar exchange rates, 4 Jan 2016 through 2 Sep 2019 (Note: The pound's daily change on 24 June 2016 is highlighted, along with its post-referendum lows. Source: Authors' calculations and landsbanki.is)

referendum level, \$1.205, on 16 January 2017, the day before Prime Minister Theresa May presented her Brexit plan. The pound-euro exchange rate fell to its post-referendum trough, €1.075, on 29 August 2017, the day before the third round of negotiations between the UK and the EU began. Shortly thereafter, the Bank of England signalled that it was preparing for an interest rate increase, which led to a sharp appreciation.[11]

Over the next few months, the pound appreciated considerably against the dollar while holding stable against the euro. The expected interest rate increase came in November 2017, and expectations of another rise in rates pushed up the value of the pound as the dollar was depreciating. The pound-dollar exchange rate peaked on 17 April 2018 but fell sharply when new data on the UK economy revealed lower-than-expected inflation, retail sales figures, and wage growth. Comments made by Bank of England Governor Mark Carney seemed to cast doubt on an expected interest rate hike in May and caused the pound to continue falling against the dollar.[12] Around this time, the dollar had begun to strengthen, causing the pound-dollar exchange rate to fall rapidly. The Bank of England delayed the interest rate rise expected in May until August, when it raised rates to 0.75%. Since then, the bank has held its bank rate unchanged, yet in the first half of 2019, the pound slowly gained ground against both the dollar and the euro. This positive trend may indicate that, as time passed, the market found that a no-deal Brexit, or hard Brexit, was becoming less likely.

On 24 May, Theresa May announced her resignation, and it soon became clear that Boris Johnson would replace her as prime minister. This triggered a surge in volatility and uncertainty, causing the pound to plunge on 3 September 2019 to \$1.196, its lowest level since 1985 (excluding the flash crash of October 2016).[13] The drop came amidst fears that Johnson would call a snap election as part of his plan to enforce a no-deal Brexit, but the pound staged a sharp recovery following the defection of Conservative MP Philip Lee, which cost the Conservatives their majority. Having lost his majority, Johnson still intends to call an early election and to suspend Parliament for a month, while MPs fight to avert a no-deal exit.

5.1 *Major Exchange Rate Movements*

We now turn to a list of major changes in the sterling exchange rate and link them to news about the UK leaving the European Union. We explore the apparent causes of

[11] See https://www.independent.co.uk/news/business/news/uk-interest-rates-latest-bank-of-england-raise-coming-months-curb-inflation-mpc-economy-a7946376.html

[12] See https://www.independent.co.uk/news/business/news/pound-sterling-latest-slide-mark-carney-may-interest-rates-bank-of-england-a8313606.html

[13] See https://www.telegraph.co.uk/business/2019/09/03/markets-live-latest-news-pound-euro-ftse-1001/

the largest daily fluctuations in the value of the pound. The standard deviation in the price of a euro and the price of the dollar was used to calculate the daily change over the entire period. The daily changes in the nominal exchange rates were then used to derive four indicator variables. The first has the value "1" if the daily change is more than double the standard deviation of the series; the second, if it is more than triple the standard deviation; and the third and fourth, if it is more than quadruple and quintuple the standard deviation, respectively. We use both sterling-euro and sterling-dollar exchange rates in order to isolate fluctuations caused solely by changes in the UK economy from those in the two trading countries.

We then compare the dates with daily changes exceeding double the standard deviation to a list of noteworthy events related to the Brexit debacle over the period from 23 January 2013 through 3 September 2019. We took the events from the House of Commons Library (2019), as well as news reports appearing in some of the UK's 'quality press' newspapers (such as *The Daily Telegraph*, the *Independent*, *The Guardian*, and the *Financial Times*), in an attempt to find the apparent cause of the exchange rate movements. This process made it possible to analyse which type (s) of news and events resulted in an appreciation of the pound and which resulted in a depreciation. Daily changes that exceeded double the standard deviation were quite common in the data. However, only 19 dates had daily changes exceeding triple the standard deviation. The figures below highlight those fluctuations, showing the exchange rates for these 4 years.

Table 2 shows the dates of the 19 daily fluctuations exceeding three times the standard deviation, the size of the change in the natural logarithm for that date, (Δlog (£/€) and Δlog(£/$)), and the most likely explanation for the significant deviation. Of the 19 large daily fluctuations, 16 are directly or indirectly related to Brexit. The largest daily swing by far took place on 24 June 2016, the date the results of the Brexit referendum were announced. In fact, sterling's drop on that date is proportionally the largest daily change in its value over the entire period from 4 January 1971 through 17 April 2019, as is shown in Fig. 14 in Appendix. The daily changes identified in Table 2 are displayed in Figs. 2, 3, 4, and 5 and highlighted. In addition to those larger swings, several daily changes apparently connected to Brexit-related events exceeded double the standard deviation. A total of 30 daily changes exceeding double the standard deviation can be traced directly or indirectly to Brexit.

5.2 A Chronology of Events

We now list the events that occurred on the days of large changes in the exchange rate, defined as days when the sterling exchange rate changed by more than three standard definitions, along with certain daily changes exceeding three standard deviations and apparently connected to Brexit-related events (Fig. 7).

Table 2 Large changes in the nominal exchange rate of pound sterling and events between 2 Jan 2015 and 3 Sep 2019

Deviation exceeds $5*\sigma^2$	Deviation exceeds $4*\sigma^2$	Deviation exceeds $3*\sigma^2$	Measured in euros/dollars	Reason for the deviation
		4.5.2015	−0.00541/−0.00599	Polls indicate the coming elections might result in a hung Parliament.
		8.5.2015	0.00797/0.00651	(7.5.2015) PM David Cameron is re-elected after promising a referendum on Brexit. He promises to fight to keep Britain in the EU and stakes his career on the outcome.
		22.1.2016	0.00594/0.00702	Mario Draghi makes a speech in Davos, which helps oil prices recover and markets stabilize after a period of turbulence.
20.6.2016			0.01256/0.01873	(16.6.2016) Labour MP, Jo Cox, is assassinated by a far-right extremist. Polls indicate increased support for Remain.
24.6.2016			−0.02577/−0.03545	The results from the Brexit referendum are announced and were as follows: *Remain*: 16,141,241 (48.1%) *Leave*: 17,410,742 (51.9%). PM David Cameron announces his resignation.
27.6.2016			−0.00966/−0.01325	Markets open for the first time post-referendum. David Cameron discusses the implications of Brexit in Parliament.
	5.7.2016		−0.00769/−0.00989	Three large investment fund managers freeze their property funds in the UK due to uncertainty regarding Brexit.
		12.7.2016	0.00659/0.00729	(11.7.2016) Confirmation that Theresa May will serve as the next PM, May did not support Brexit.
		4.8.2016	−0.00512/−0.00623	The Bank of England lowers interest rates in response to Brexit.
	7.10.2016		−0.00850/−0.00797	Temporary turbulence in Asian markets causes a sterling "flash crash".
		9.1.2017	−0.00562/−0.00674	(8.1.2017) In an interview, PM Theresa May suggests that she aims for a "hard Brexit".

(continued)

Table 2 (continued)

Deviation exceeds $5*\sigma^2$	Deviation exceeds $4*\sigma^2$	Deviation exceeds $3*\sigma^2$	Measured in euros/dollars	Reason for the deviation
	17.1.2017		0.00795/0.01173	Theresa May declares that any agreement on leaving the EU will have to be ratified by parliament.
		18.4.2017	0.00513/0.00894	Theresa May announces snap elections.
		9.6.2017	−0.00606/−0.00711	(8.6.2017) The Conservative Party loses its majority in national elections but manages to form a new government.
		9.2.2018	−0.00612/−0.00624	Michel Barnier, European Chief Negotiator for the United Kingdom Exiting the European Union, claims that a "no-deal" Brexit is a possible outcome.
		21.9.2018	−0.00581/−0.00627	Theresa May makes a speech and says that there has been little progress in the negotiations with the EU.
		15.11.2018	−0.00725/−0.00742	(14.11.2018) The agreement with the EU is published and meets objections both within and outside the Conservative Party; (15.11.2018) Ministers resign to protest the agreement.
		10.12.2018	−0.00704/−0.00687	May postpones voting on the Brexit agreement.
		29.7.2019	−0.00667/−0.00635	Boris Johnson's new government takes a tough stance on a no-deal Brexit, with Michael Gove writing in the Sunday Times that the government was now "working on the assumption" of a no-deal Brexit.

Source: Authors' calculations, British newspapers, and House of Commons Library (2019)
Note: The standard deviation over the period was: $\sigma^2 = 0.00171$ measured in euros, and $\sigma^2 = 0.00196$ measured in US dollars

4 May 2015

Capping a 3-day continuous fall in the value of the pound, its sharp devaluation on 4 May 2015 of more than three standard deviations reflected traders' fears over a possible hung Parliament in the upcoming general election. Although Brexit was already a salient political issue at that point, we do not regard this as a Brexit-related event.

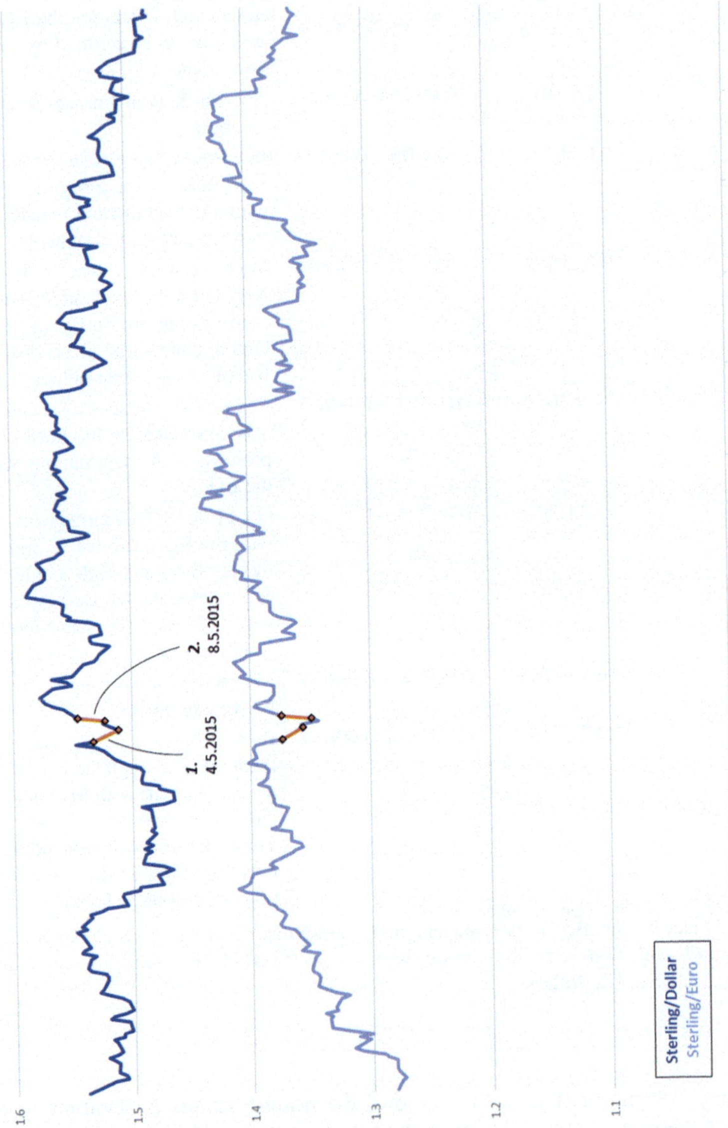

Fig. 7 Pound-euro and pound-dollar exchange rates in 2015 (Note: The two daily changes exceeding triple the standard deviation are highlighted. Source: Authors' calculations and Landsbankinn)

8 May 2015

On the evening of 8 May 2015, then-current Electoral Commission Chair Jenny Watson announced the results of the UK's general election, with the Conservative Party winning a clear majority and David Cameron remaining prime minister. The Conservative Party had promised a referendum on EU membership, a promise that David Cameron reiterated in his victory speech, while also stating his determination to fight against an exit. The pound appreciates, but again, this is not a Brexit-related event (Fig. 8).

22 January 2016

Recovering oil prices and market optimism following a speech by Mario Draghi caused an appreciation exceeding triple the standard deviation.

20 June 2016

On Thursday 16 June, Labour Party MP Jo Cox was assassinated by a far-right fanatic. It was the first murder of a sitting MP in the UK since 1990, and all campaigning ceased. Polls taken on 20 June indicated growing support for Remain, leading to a sharp appreciation (more than five standard deviations) of the pound sterling, the second-largest daily change in the entire period. This is our first Brexit-related event.

24 June 2016

On Friday 24 June, the results of the Brexit referendum were announced.[14] That same morning, Prime Minister David Cameron announced his resignation outside 10 Downing Street. The results came as a shock to markets, in contrast to the notion of the wisdom of crowds,[15] and while Nigel Farage and other Brexiteers celebrated, the pound plummeted. The daily change in the pound's value was proportionally the largest in the history of its free-floating exchange rate. This is our second Brexit-related event.

27 June 2016

Markets reopen on Monday 27 June and resume their free fall. The daily change in the exchange rate exceeds five times the standard deviation and is the third-largest over the period in question. Again, this is related to Brexit.

29 June 2016

David Cameron urges the EU and the UK to be optimistic and show solidarity in his first post-referendum speech in the European Parliament. The pound slightly recovers after the chaos of the previous days. Although the appreciation is small in comparison to previous losses, just exceeding double the standard deviation, it indicates that initial reactions may have been overblown. Brexit-related.

[14] They were as follows: *Remain*: 16,141,241 (48.1%) *Leave*: 17,410,742 (51.9%).

[15] See Biggs (2008).

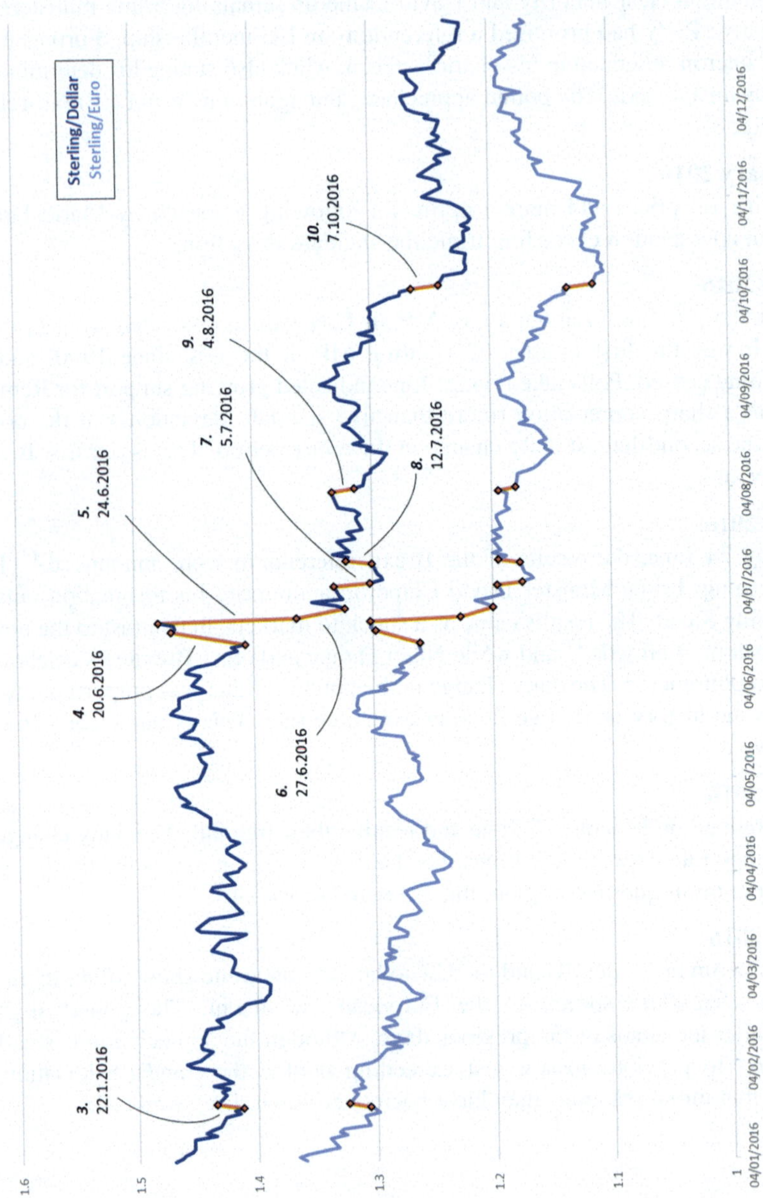

Fig. 8 Pound-Euro and Pound-Dollar exchange rates in 2016 (Source: Authors' calculations and Landsbankinn)

30 June 2016
Mark Carney gives a speech on Britain's future outside the EU. His expectations are gloomy and pessimistic, and he foresees that weaker growth and higher inflation could be in the offing (Carney 2016). Following the speech, the pound depreciates by more than double the standard deviation, erasing the appreciation from the day before. Brexit-related.

5 July 2016
Three large fund managers freeze their property funds in the UK due to uncertainty surrounding Brexit. A study showing slowing growth in the UK's service sector is released on the same day, resulting in a depreciation of more than four times the standard deviation. By the end of the day, the pound has sunk to a 31-year low against the dollar.[16] Brexit-related.

12 July 2016
On 11 July, Theresa May's appointment as leader of the governing Conservative Party and as prime minister is confirmed. Boris Johnson, former mayor of London and one of Farage's leading supporters in the Leave campaign, had formerly been regarded as a likely candidate. May had been opposed to Brexit, so her appointment inspires those hoping for a soft withdrawal. In her victory speech, May nonetheless promises her party colleagues that no attempt will be made to overturn the referendum's results, saying "Brexit means Brexit." The pound reacts to the day's events with an appreciation exceeding triple the standard deviation.[17] Brexit-related.

14 July 2016
Theresa May is formally appointed prime minister. The pound appreciates more than double the standard deviation. The Brexit process officially begins.

4 August 2016
The Bank of England cuts its interest rates for the first time in more than 7 years, in measures designed to prevent a post-Brexit recession. The cut, from 0.5 to 0.25%, is part of the bank's four-point stimulus package meant to deal with post-Brexit economic fallout. The pound's tumble, exceeding triple the standard deviation, indicates that the bank surprised markets with its response to an expected economic slowdown following the vote.

3 October 2016
In her Party Conference speech, May introduces "The Great Repeal Bill" repealing the European Communities Act 1972, which gives legal authority to EU law within the UK. The bill would require that hundreds of EU laws be abolished and replaced with new British laws, without going through primary legislation. In the same speech, she confirms that she will invoke Article 50 of the Treaty on European Union before the end of March 2017, declaring that she would not require the

[16]See https://www.theguardian.com/business/2016/jul/05/pound-hits-31-year-low-after-service-sector-data

[17]See https://www.ft.com/content/ef6d5a38-4920-11e6-b387-64ab0a67014c

support of the House of Commons to do so. The House of Lords criticises the bill as anti-democratic. The pound responds with a depreciation exceeding double the standard deviation.[18] Brexit-related.

7 October 2016
In a two-minute flash crash, the pound sterling drops more than 6% against the US dollar, and although it regains most of its losses before the end of the day, the daily depreciation still exceeds four times the standard deviation. The flood was initially believed to have been caused by a fat-finger trading error, but there were indications that it may have been a reaction to news of French President François Hollande's tough stance over Brexit negotiations.[19]

18 October 2016
The High Court of Justice hears the Gina Miller case challenging Theresa May's statements that the UK government would not require the approval of Parliament to trigger Article 50. The pound appreciates more than double the standard deviation. Brexit-related (Fig. 9).

9 January 2017
In her New Year's television interview, Theresa May suggests that the UK is heading towards a "hard Brexit" and says that Britain "will not attempt to cling on to bits of EU membership."[20] Predictably, currency markets react poorly and the pound falls, with the depreciation exceeding triple the standard deviation. Brexit-related.

17 January 2017
In a speech in Lancaster House, May emphasises her intention for the UK to leave the EU's internal market but does not give definite answers regarding the customs union. Moreover, she confirms for the first time that the government will subject any deal agreed between the UK and the EU to a vote in Parliament,[21] making it clear that the government would not force through a hard Brexit without Parliament's approval. The speech results in an appreciation exceeding four times the standard deviation. Brexit-related.

25 January 2017
On 24 January, the Supreme Court confirms the High Court's ruling that the government requires approval from both Houses of Parliament approval to trigger Article 50. Sterling appreciates more than double the standard deviation. Brexit-related.

---

[18]See https://www.independent.co.uk/news/uk/politics/great-repeal-bill-brexit-law-eu-law-theresa-may-david-davis-a7343256.html

[19]See https://www.ft.com/content/dfb375be-8c23-11e6-8cb7-e7ada1d123b1

[20]See https://www.independent.co.uk/news/business/news/pound-sterling-value-brexit-leaving-single-market-low-theresa-may-interview-a7516766.html

[21]See https://www.theguardian.com/politics/2017/jan/17/key-points-from-mays-what-have-we-learned

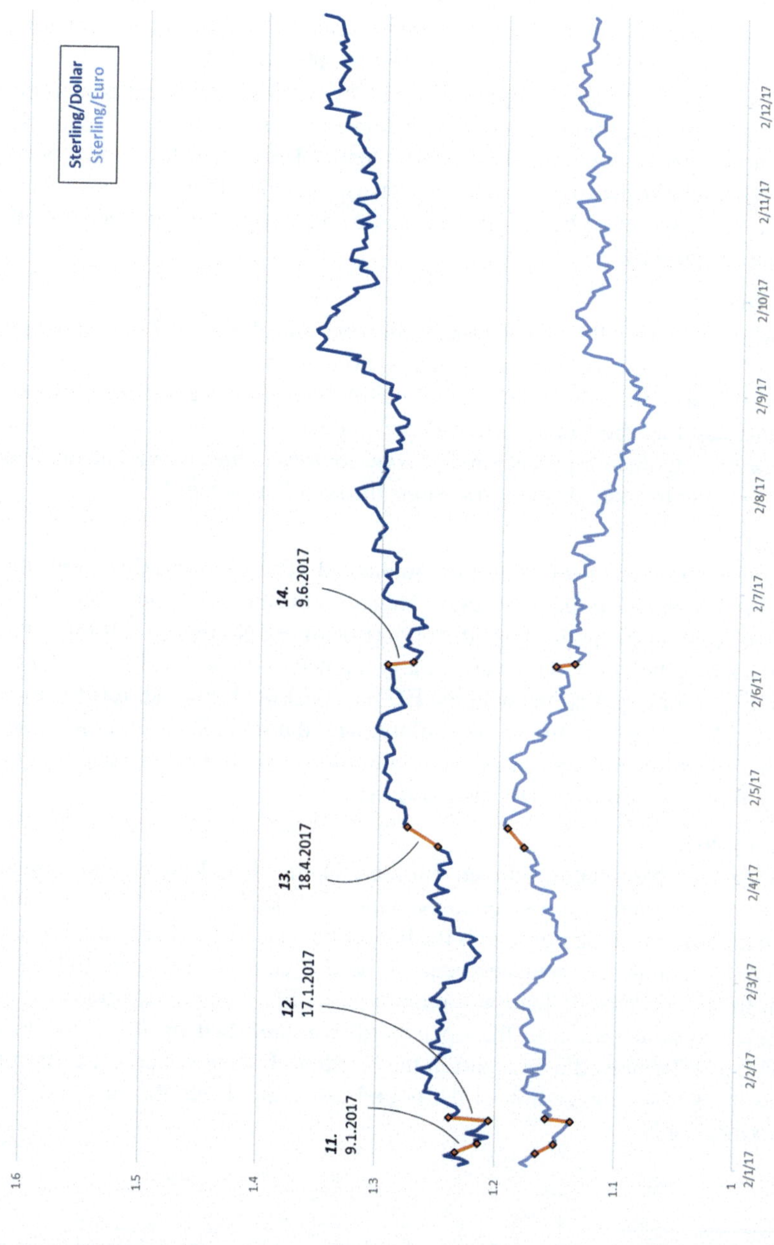

Fig. 9 Pound-euro and pound-dollar exchange rates in 2017 (Source: Authors' calculations and Landsbankinn)

8 February 2017
The European Union Bill passes its third reading in the House of Commons, from where it will move on to the House of Lords. The bill permits Theresa May's government to trigger Article 50. The pound reacts by appreciating by more than double the standard deviation. This is the only instance of a positive reaction, albeit a minor one, to an event indicating an increased likelihood of Brexit.

The House of Commons' approval could be seen as an event that reduces uncertainty. From this point forward, Brexit-related news and the currency market's reactions no longer revolve mainly around whether Brexit will happen or not, but rather whether a hard Brexit, or a no-deal scenario, is more likely or less so. For the markets, it would therefore be best if negotiations were to run swiftly and smoothly, minimising uncertainty.

18 April 2017
Theresa May unexpectedly calls a general election on 8 June 2017 in an attempt to strengthen her position for upcoming negotiations with the EU. The pound reacts positively, its gains exceeding triple the standard deviation. The appreciation comes as polls indicate that the Conservative Party could win by a large majority, which would ease the negotiation process and prevent long-term uncertainty, as the majority would be able to force difficult decisions through Parliament.[22]

9 June 2017
The results of the general election are announced. The Conservative Party unexpectedly loses its majority in the election but still wins the most seats. Theresa May manages to form a new majority with support from the Northern Irish DUP Party. Her plan to strengthen the Conservative Party's position domestically, in order to be better equipped for negotiations with the EU, has backfired. Instead, results reveal a national divide adding to political uncertainty, with the small DUP Party agreeing to defend the government subject to certain conditions. Sterling depreciates by more than three times the standard deviation (Fig. 10).

9 February 2018
The EU's Chief Negotiator, Michel Barnier, claims that May's government is unwilling to compromise on its demands, which will lead to a no-deal exit with no transition period. The main obstacle is the border between the UK and Ireland: a hard Brexit would require the reinstatement of border controls between Ireland and Northern Ireland because Ireland will remain in the EU. Such a hard border is very undesirable, however, because the open borders established by the Good Friday Agreement in 1998 played a vital part in the Northern Irish peace process. Barnier's statement leads to a depreciation of the pound exceeding triple the standard deviation. Brexit-related.

[22]See https://www.theguardian.com/business/live/2017/apr/19/ftse-100-election-pound-market-imf-business-live?page=with:block-58f737d6e4b05776df18f0ca#block-58f737d6e4b05776df18f0ca

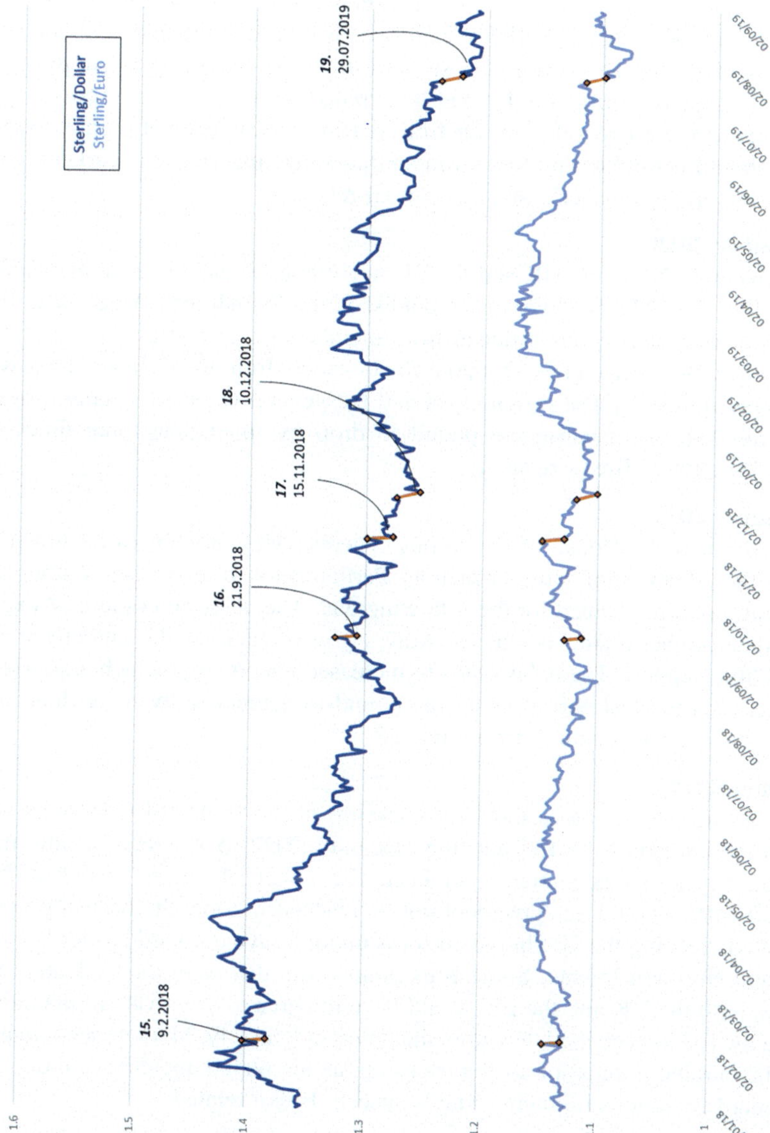

Fig. 10 Pound-euro and pound-dollar exchange rates, 2 Jan 2018 through 3 Sep 2019 (Source: Authors' calculations and Landsbankinn)

21 September 2018
Theresa May releases a statement on two issues that had stood in the way of an agreement. The former concerns the post-Brexit economic relationship between the EU and the UK. The EU wants the UK either to remain part of the internal market and customs union or to set up border controls between the EU and the UK, but May finds neither option acceptable. The second issue is related to the first, as it concerns the Irish border and the so-called Irish backstop. The backstop is supposed to guarantee that a closed Irish border will be avoided, no matter what. According to May's statement, the two sides are far from reaching an agreement on how such a backstop should work. Sterling falls by more than three times the standard deviation following the prime minister's remarks. Brexit-related.

15 November 2018
On 14 November 2018, the UK and the EU negotiating teams reach an agreement, and later that day, May's government publishes the Withdrawal Agreement. The agreement is met with fierce criticism from members of all parties. The next day, 15 November, Brexit Secretary Dominic Raab resigns from the Cabinet along with several other ministers. The resignations and the general negative reception of the Withdrawal Agreement cause the pound to drop by more than three times the standard deviation.[23] Brexit-related.

10 December 2018
In a statement to the House of Commons, Theresa May announces an indefinite delay in the Meaningful Vote, which is Parliament's vote on her Withdrawal Agreement, initially planned for the following day. The delay comes amidst expectations of an embarrassing defeat for May's government, as the agreement had garnered little support. The delay leads to increased uncertainty, which exacerbates the danger of a no-deal exit, causing the pound to depreciate by more than three times the standard deviation.[24] Brexit-related.

14 January 2019
The prime minister discusses the Irish backstop in the House of Commons. The backstop, which gives the EU and the UK until 2022 to negotiate a new trade agreement before its activation, had been the target of severe criticism from pro-Brexit MPs. Should an agreement not be reached in time, the backstop would be activated, keeping the UK in the customs union to avoid a hard border between Ireland and Northern Ireland. Some MPs pointed out that were the backstop to be activated, both the UK and the EU would have to agree to deactivate it, potentially allowing the EU to keep the UK within the union indefinitely. Markets seem to have a more favourable impression of the backstop, as the pound appreciates more than two standard deviations following May's speech. Brexit related.

---

[23]See https://www.theguardian.com/business/2018/nov/15/pound-falls-steeply-as-brexit-resignations-rock-the-government

[24]https://www.telegraph.co.uk/business/2018/12/11/value-pound-has-affected-theresa-may/

26 February 2019

Theresa May promises the House of Commons a vote to rule out a no-deal Brexit on 29 March should she lose the second Meaningful Vote next month. As expected, the pound reacts positively, gaining more than double the standard deviation over the day. Brexit-related.

22 March 2019

The European Council meets in Brussels and agrees to an extension of Article 50 until 22 May 2019 if May's third Withdrawal Agreement is approved by the House of Commons, or until 12 April 2019 if it is not. Because the original exit date, 29 March, is only a week away, the extension is essential to avoid a no-deal Brexit, and predictably, the pound appreciates following the news. The pound-euro exchange rate rises by more than four times the standard deviation and the dollar-euro exchange rate by more than double the standard deviation. Brexit-related.

3 May 2019

The Conservative Party loses over 1000 councillors and control of several councils in local elections. Labour fails to gain from these losses, instead suffering some losses of its own, while the Liberal Democrats, Greens, and Independents fare well. The pound reacts poorly, depreciating more than double the standard deviation over the day. This was the beginning of a slide from which it has yet to recover, indicating growing fears of a no-deal Brexit. Brexit-related.

29 July 2019

Boris Johnson's new government takes a tough stance on a no-deal Brexit, with Michael Gove writing in the Sunday *Times* that the government was now "working on the assumption" of a no-deal Brexit. While the prime minister himself doesn't claim that no-deal is the working assumption, his deputy official spokesperson does, as well as making it clear that he will not be meeting with EU leaders until an Irish backstop is off the table.[25] Markets reacted with fear, causing the pound to depreciate more than three standard deviations. Brexit-related.

5.3 *Summary of Effects on the Exchange Rate*

Judging from these swings in the pound's exchange rate in response to Brexit-related news, one can conclude that markets have yet to foresee the nature of the UK's exit from the EU and that sterling tends to fall with both increased uncertainty and increased likelihood of a hard exit. The Brexit risk may raise the currency premium or the country premium.

[25] See https://www.theguardian.com/business/2019/jul/29/pound-drops-lowest-level-in-two-years-amid-no-deal-brexit-rhetoric-sterling-dollar-euro

Interestingly, the more time has passed from the Brexit referendum, the less impact specific news or reports on the state of negotiations seem to have on the pound. It is possible that the uncertainty has become so pronounced that the words of politicians have limited impact. If so, markets might avoid any gambles and wait until a Withdrawal Agreement passes through Parliament or a no-deal exit takes place. It could also be that the market had indeed foreseen most of the more recent developments—i.e., that Theresa May would lose the third Meaningful Vote—and had already taken them into account, limiting large swings. Even now, in the face of Johnson's strong no-deal rhetoric, markets still seem hesitant to price in the implications of such an exit, although the new prime minister's tenure in office has been followed by a steady, if slow, slide in the value of the pound.

6 Conclusions

The Brexit vote has primarily affected the sterling exchange rate. The more likely the UK is to leave the EU without an agreement, the further the exchange rate falls. The gradual rise in the exchange rate in recent years would then tell us that the currency market is leaning away from a hard Brexit outcome. There is also some effect on the FTSE 250, but this is more muted. The effect on the real economy is weak; employment is booming and capital inflows continue, although investment appears to be falling and house price inflation has lost pace.

It is clear that there is much uncertainty about future developments, and that this uncertainty dampens the response of markets and the economy. From what we can see, a hard Brexit would have a negative impact on the exchange rate, the stock market, and output. But the effect would be mitigated, even neutralised, if an agreement were reached that ensured continued free trade between the UK and the Continent. Such an agreement might even raise the exchange rate and output from current levels by eliminating uncertainty.

Appendix

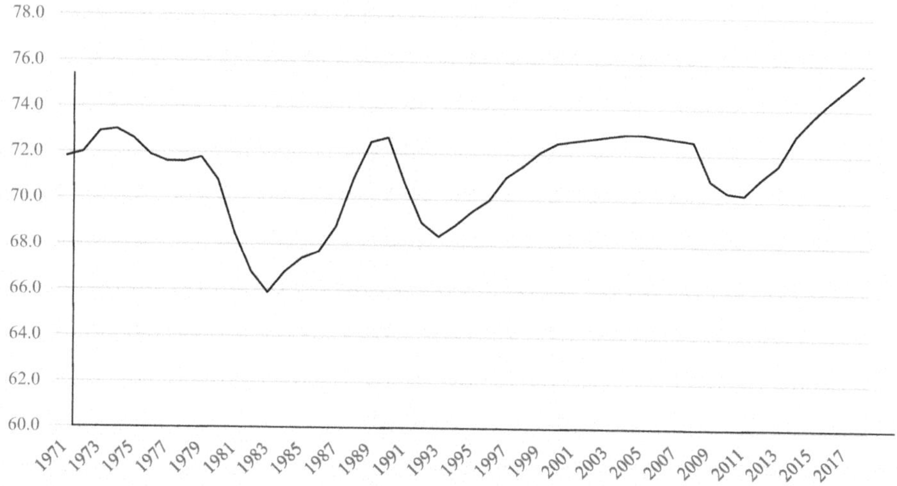

Fig. 11 Employment as a share of the population aged 16–64 (%) (Source: Office for National Statistics)

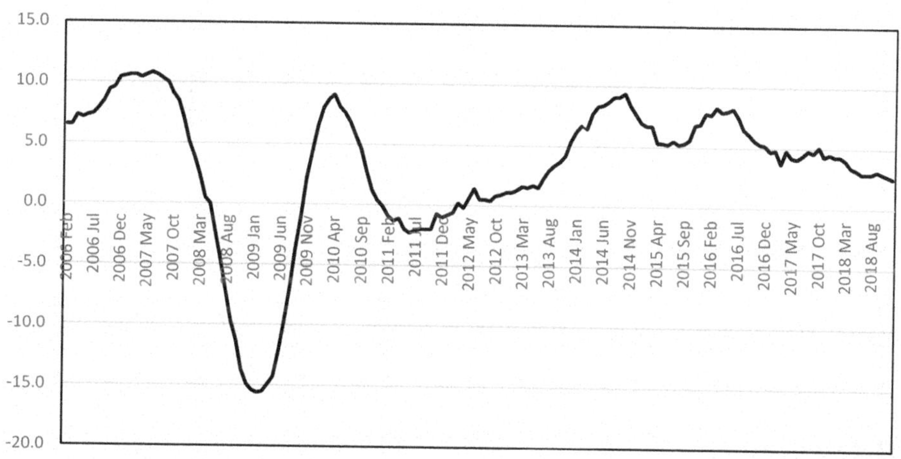

Fig. 12 Annual rate of growth of UK house prices (%)

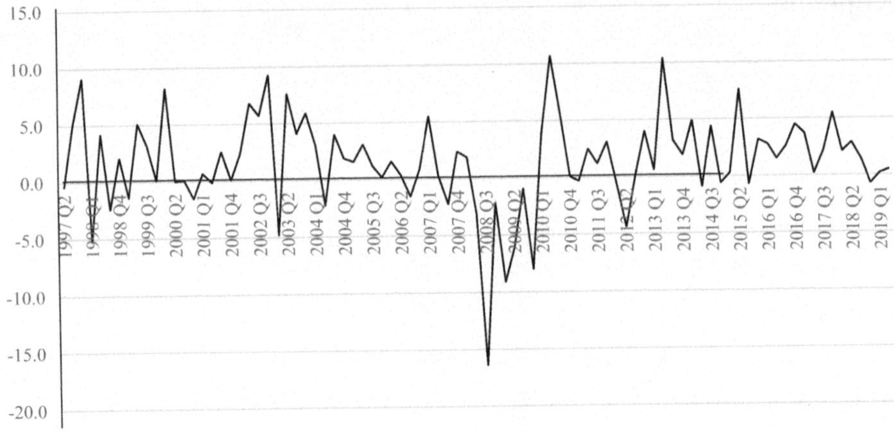

Fig. 13 Gross fixed capital formation—quarterly changes (%) (Source: Office for National Statistics)

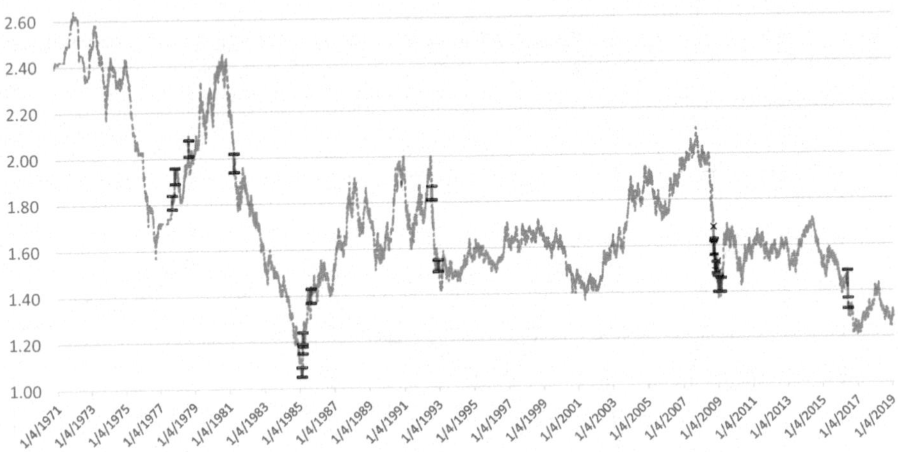

Fig. 14 The pound-dollar exchange rate, 1971–2019 (Source: Authors' calculations and investing.com)

References

Antonucci, L., Horvath, L., Kutiyski, Y., & Krouwel, A. (2017). The malaise of the squeezed middle: Challenging the narrative of the 'left behind' Brexiter. *Competition and Change, 21*(3), 211–229.

Arnorsson, A., & Zoega, G. (2019). On the causes of Brexit. *European Journal of Political Economy, 55*, 301–323.

Biggs, B. (2008). *Wealth, war & wisdom.* Hoboken: Wiley.

Carney, M. (2016, June 30). Uncertainty, the economy and policy. *Speech by Mark Carney*. Retrieved April 29, 2019, from https://www.bankofengland.co.uk/speech/2016/uncertainty-the-economy-and-policy

Colantone, I., & Stanig, P. (2018). Global competition and Brexit. *American Political Science Review, 112*(2), 201–218. https://doi.org/10.1017/S0003055417000685.

Corsetti, G., & Müller, G. (2016). The pound and the macroeconomic effects of Brexit. *Vox EU*. Retrieved April 20, from https://voxeu.org/article/pound-and-macroeconomic-effects-brexit

Crafts, N. (2019, January 15). Brexit: Blame it on the banking crisis. *Vox EU*. Retrieved April 12, 2019, from https://voxeu.org/article/brexit-blame-it-banking-crisis

Davies, R. B., & Studnicka, Z. (2018). The heterogeneous impact of Brexit: Early indications from the FTSE. *European Economic Review, 110*, 1–17. https://doi.org/10.1016/j.euroecorev.2018.08.003.

Dhingra, S., Ottaviano, G. I. P., Sampson, T., & Reenen, J. (2016). *The consequences of Brexit for UK trade and living standards*. CEP BREXIT Analysis No. 2.

Dustman, C., & Frattini, T. (2014). The fiscal effects of immigration to the UK. *The Economic Journal, 124*(580), F593–F643.

Fidrmuc, J., Hulényi, M., & Tunalı, Ç. (2016). *Money can't buy EU love: European funds and the Brexit referendum*. CESifo Working Paper Series No. 6107.

Forte, G., & Portes, J. (2017). The economic impact of Brexit-induced reductions in migration. *Oxford Review of Economic Policy, 33*(1), S31–S44. https://doi.org/10.1093/oxrep/grx008.

Goodwin, M. J., & Heath, O. (2016). The 2016 referendum, Brexit and the left behind: An aggregate level analysis of the result. *The Political Quarterly, 87*, 323–332. https://doi.org/10.1111/1467-923X.12285.

Hill, C. A. (1997). How investors react to political risk symposium. International Issues in Cross-Border Securitization and Structured Finance, Duke Journal of Comparative & International Law.

Hobolt, S. B. (2016). The Brexit vote: A divided nation, a divided continent. *Journal of European Public Policy, 23*(9), 1259–1277. https://doi.org/10.1080/13501763.2016.1225785.

House of Commons Library. (2019). https://commonslibrary.parliament.uk/tag/nigel-walker/.

Inglehart, R., & Norris, P. (2016). *Trump, Brexit and the rise of populism: Economic have-nots and cultural backlash*. HKS Working Paper No. RWP16-026. https://doi.org/10.2139/ssrn.2818659

Inglehart, R., & Norris, P. (2019). *Cultural backlash: Trump, Brexit, and authoritarian populism*. Cambridge: Cambridge University Press.

Knight, F. H. (1921). *Risk, uncertainty, and profit*. Boston and New York: Houghton Mifflin Company.

Mudde, C. (2004). The populist zeitgeist. *Government and Opposition, 39*(4), 541–563. https://doi.org/10.1111/j.1477-7053.2004.00135.x.

Mudde, C. (2017). *Populism: A very short introduction* (2nd ed.). New York: Oxford University Press.

Müller, J.-W. (2015). Parsing populism: Who is and who is not a populist these days? *Juncture, 2*(22), 80–89. https://doi.org/10.1111/j.2050-5876.2015.00842.x.

Müller, J.-W. (2016). *What is populism*. Philadelphia: University of Pennsylvania Press.

Rodrik, D. (2018). Populism and the economics of globalization. *Journal of International Business Policy, 1*(1–2), 12–33. https://doi.org/10.1057/s42214-018-0001-4.

Wright, B. (2016, June 28). Why we should be looking at the FTSE 250 and not the FTSE 100 to gauge the impact of Brexit. *The Telegraph*. Retrieved May 2, 2019, from https://www.telegraph.co.uk/business/2016/06/27/why-we-should-be-looking-at-the-ftse-250-and-not-the-ftse-100-to/

Climate Change: A New European Union Approach Is Needed

Tullio Fanelli and Alessandro Ortis

Abstract The Emission Trading Scheme (ETS), on which the European Union (EU) targets to mitigate climate change are largely based, has proved to be inefficient, especially at counteracting the effects of global trade growth. In fact, although the replacement of European products with emerging Countries' imports has led to an apparent decrease in EU's emissions, it has, at the same time, brought about a substantial increase in global emissions, because of the lower efficiency and higher environmental impacts of most imported products. The risk is that the ETS, which was developed on a territorial basis, while unable to provide substantial benefits to the environment, proves to be an additional factor for CO_2 increases, and the loss of competitiveness of EU's industry, that is exposed to international environmental dumping. The strategy for Europe to set their irrefutable environmental targets, while protecting their manufacturing industry, must be mainly based on a system of emission tracking and specific labelling that acknowledges the higher environmental quality of the European products, thus enabling consumers to choose the most eco-friendly ones. Furthermore, a privileged taxation could replace the ETS system. It must be a non-discriminatory, non-protectionist taxation, that simply offsets the economic advantage of environmentally impacting products over eco-friendly ones.

Keywords Environmental · Countries · Global · Goods · Services · Tax

T. Fanelli (✉)
ENEA, Italian National Agency for New Technology, Energy and Sustainable Economic Development, Rome, Italy
e-mail: tullio.fanelli@enea.it

A. Ortis
ENEA's Scientific and Technical Committee, Rome, Italy
e-mail: a.ortis@ortis.onmicrosoft.com

L. Paganetto (ed.), *Capitalism, Global Change and Sustainable Development*, Springer Proceedings in Business and Economics,
https://doi.org/10.1007/978-3-030-46143-0_9

1 Introduction

Although at the Paris climate conference (COP21) in December 2015, 195 Countries agreed on the long-term goal of keeping global average temperature increase below 2 °C, compared to pre-industrial levels, global emissions actually continue to rise (Fig. 1).

As is known, the EU is the most active actor in the field: their Nationally Determined Contribution (NDC) under the Paris Agreement is to reduce greenhouse gas emissions by at least 40% compared to 1990, by 2030, as defined in the EU's wider 2030 climate and energy framework. Nonetheless, the creation of a global carbon market on the model of the EU's *cap and trade* ETS system seems extremely unlikely; the ETS system risks remaining a non-unified model, applied in Europe and in hardly few other Countries, while serious doubts are starting to be raised as to the efficacy of its objectives and the adequacy of its instruments. Actually, it must be considered that, over the past 20 years, significant changes, related to the huge expansion in the world trade have occurred on a global scale. This has contributed to determine a significant increase in emissions of exporting Countries that still have to adopt appropriate measures to fight climate change (Fig. 2).

As an example, after 2001, when they obtained the World Trade Organization (WTO) membership, China tripled their emissions, thus achieving almost 10 CO_2

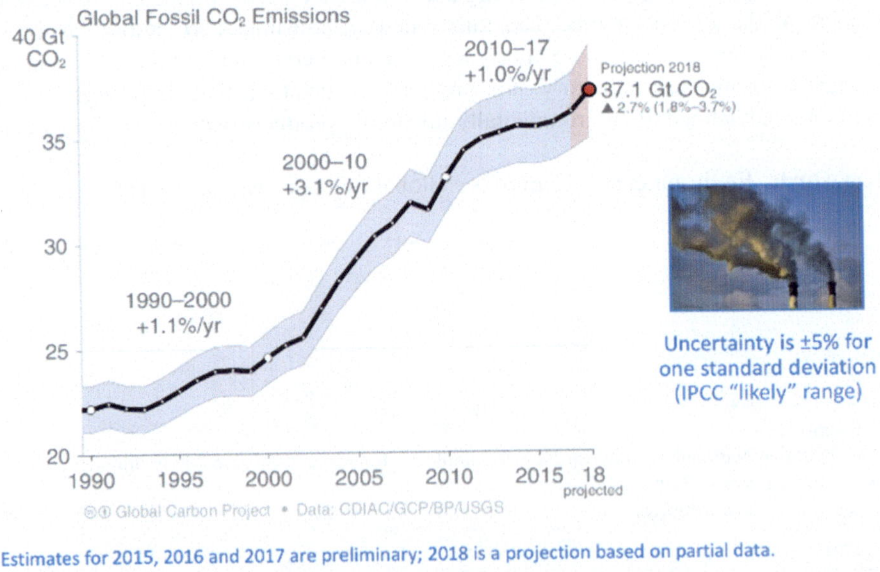

Fig. 1 Global fossil CO_2 emissions: 36.2 Gt CO_2 in 2017, 63% over1990—projection for 2018: 37.1 ± 2 Gt CO_2, 2.7 higher than 2017 (range 1.8–3.7%) (Source: CDIAC)

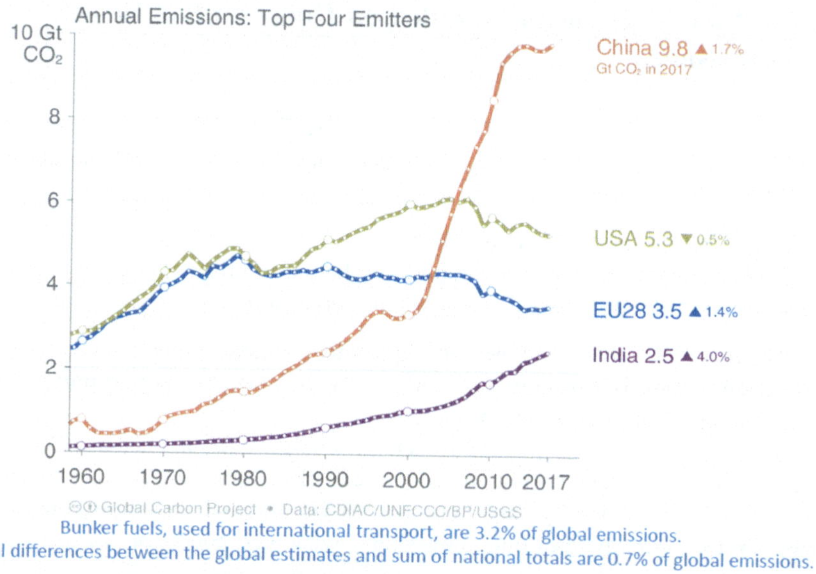

Fig. 2 the top four emitters in 2017 covered 58% of global emissions: China (27%), United States (15%), EU (10%), India (7%) (Source: CDIAC)

equivalent tons (t CO_2) in 2017, i.e. more than the overall sum of the emissions produced in the USA (5.3 t CO_2) and the UE (3.5 t CO_2).

These data, along with those from other Countries with a high growth rate of net exports, suggest that emissions from international trading activities have become crucial to understand the change of global emission patterns.

Today, the parallel occurrence, in a given geographical area, of emissions derived from production processes (*production based*) and those derived from consumption (*consumption based*) of the goods and services produced, is no longer a straightforward fact.

The emission accounting model adopted worldwide, which was exclusively developed on a territorial basis, cannot take these new phenomena into account: the decrease in emissions in some Country is not actually a guarantee of some corresponding reduction in world emissions.

Macroeconomic analyses and recent estimates in the literature emphasize how significant emissions from the consumption of goods and services (*consumption based*) that are mainly associated with imports from the major emerging economies to western Countries, are.

Should greenhouse gas emissions accounting be reconsidered on the basis of consumption responsibility rather than on emission territoriality, dramatically different results could ensue.

2 A New Mechanism Founded on Consumption Based Emissions

At this point a serious reflection upon the responsibility for the actual global emission increase is necessary, as well as on new approaches and measures which should:

- ensure more highly effective objectives;
- entail greater consumers' awareness on the environmental impacts of their consumption behavior, regardless of where goods and services are produced.

The major difficulty of implementing a new mechanism founded on consumption based emissions is certainly the complexity of correctly apportioning emission accountability for goods and services.

Yet a new EU energy taxation directive, providing for an excise tax component, proportional to induced CO_2 emissions, can be a first significant step ahead to overcome such difficulty and approach environmental accounting with a new method for properly ascribing emissions, whatever goods or services they are derived from.

Regardless of its quantitative dimension, such tax component would indeed allow to calculate the emissions related to production activities with a simple ratio. In fact, the emissions due to a single activity would be equal to the ratio between:

- the sum (€) of all tax components paid (relating to each energy sources), and
- the coefficient of proportionality between the tax component and CO_2 emissions (€/t CO_2).

If only one kind of product (goods or service) is produced, then, of course, the emissions are equal to the ratio between total CO_2 emitted during production and the produced units.

If two or more product types are produced, total emissions due to production activities are to be allotted among the types, by assessing their share of energy consumption.

If CO_2 emission information is transferred along the entire production and distribution chain as the billing is, it would then be possible to easily track goods and services' emissions with no significant administrative burden, at least when they are entirely produced in Europe.

Figure 3 is a chart representing such mechanism: the flow of emissions in a simplified production chain is shown.

The total amount of induced emissions could then be reported on product labels, together with the price and other useful information on product characteristics. This would already be an important result per se, as it would allow:

- consumers to steer their own choices also according to the lower, or higher, carbon content of the goods they are to purchase;
- companies to implement eco-friendly marketing policies.

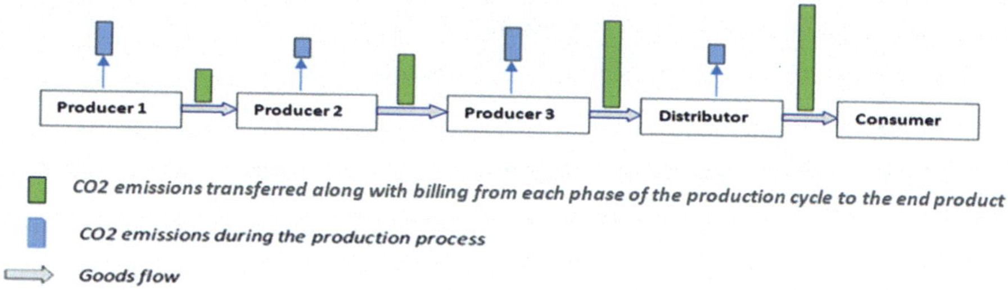

Fig. 3 Emissions induced mechanism for consumer goods

Of course, attributing given amounts of emissions to goods and services is conducive to new instruments to curb those very emissions. Specifically, a new and innovative tax system which takes into account the carbon content of goods and services, could be devised.

For example, an indirect tax, the *Carbon-Added Tax* (CAT), could be introduced: a mechanism similar to the Value Added Tax, which is an additional cost, not for the production chain operators, but rather for the end consumers.

The necessary steps for collecting such a tax could be summarized as follows:

- the non-domestic use of fuels, propellants and electric power would be additionally taxed (CAT) in proportion to the CO_2 emitted;
- the sellers of the above energy products would pay the relative CAT to the Treasury Department;
- the producers of end or intermediate goods and services would repay their suppliers (whether of energy products, raw materials, or intermediate goods and services) of the paid CAT, separately stating its amount on the invoice;
- the end seller, having declared both the associated emissions and the related tax amounts on the invoice or bill, would get the CAT of the whole production chain paid back from their customers.

Figure 4 is a chart representing the new taxation mechanism, where the flows of emissions, money and goods in a simplified production chain are shown.

The aim of such innovative tax system would be to provide consumers with an additional price signal, some extra information that is usually not stated on environmental labels; the signal could vary accordingly to the emission reduction targets to be achieved. What is more, such a mechanism could also positively impact on the production chain, since when operators made a purchase, whether a semi-worked or an end product, they would directly acquire the information conveyed by the CAT price, and would be given the chance to steer their own choices towards environmentally sustainable products, which should ensure higher competitiveness.

The CAT could be complementary to the VAT, a form of worldwide spread indirect tax, with a neutral transaction, that is, it should actually not add up to the total of indirect taxation. Such a target could be achieved by reducing the VAT rate (of all goods and services) by a percentage closely equivalent to the revenue expected from the new CAT tax.

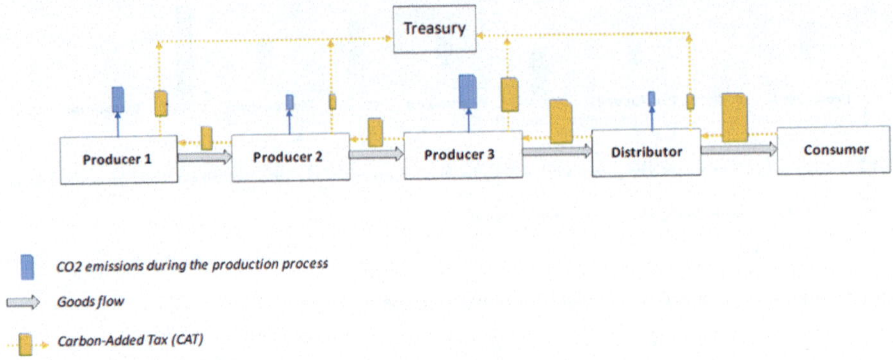

Fig. 4 Emission-based taxation mechanism for consumer goods

Alternatively, the CAT could be totally or partially conceived as an additional tax to get resources for funding climate change mitigation and adaptation measures: R&D of innovative technologies, technology transfer towards the world poorest Countries, etc.

The CAT could be directly proportional to the amount of product-related emissions or differentiated according to the deviation from an average, or *benchmark* value, of the intensity of emissions. The *virtuous* products could thus be even more incentivized with partial or total tax exemptions, counterbalanced by the effect of higher rates applied to those products most deviating from the reference value. On equal tax pressure conditions, the best low-emission production technologies would benefit from an even faster intervention. Otherwise, the financial flows allowing *virtuous* producers to get back the CAT paid to their suppliers would be more complex.

At first, the new system could be applied only in the EU without infringing the rules of the WTO: end or intermediate products exported outside the EU could benefit from a CAT exemption, whereas the CAT could be imposed to EU imported goods, according to transparent and non-discriminating criteria. For instance, the amount of a CAT on a given imported product could be calculated totaling:

- the average amount of the emissions specifically referred to the same production in the EU;
- the different emission rate per unit of GDP of the Country of origin of the product;
- the emissions related to intercontinental transport.

Yet, such a model would prove attractive both to developed Countries (other than the EU's) and emerging economies, which could be led to adopt similar tracking mechanisms for emissions induced by exported goods and services, although to merely benefit from the tax breaks provided for in the European market.

The potential advantages of this new approach to greenhouse gas emissions reduction are therefore manifold. Actually, the new mechanism could:

- allow to better exploit eco-friendly goods and services regardless of the country of origin, thus triggering a competition process among emerging Countries for more environmentally efficient productions;

- implement border levelling of the costs that depend on the environmental factors associated to imported and exported goods, bypassing the risk attributable to the ETS system of influencing the decisions on the localization of high emission intensity plants; (obviously, such an effect would be negligible for those sectors where decisions on localization are mainly taken on the basis of different factors, e.g. labor cost, environmental and safety regulations, etc.);
- help keep track of the emissions induced by goods or services, with the consequent advantage of greater awareness among end consumers, who are given the chance to choose whatever goods, not merely considering their price, but also their impact on the environment.

It is worth noting that the new taxation mechanism is radically different from the traditional proposals to introduce a *carbon tax* (a solution which can make calculations easier but whose global scale adoption is politically hard to realize); actually, levying non-deductible excise taxes on energy products, although impacting on the prices of goods and services, is often just a blur to consumers, who cannot grasp the environmental advantage of the measure, among many others tax measures that make prices increase all the same. Besides, a *carbon tax* one-sidedly introduced in the EU would cause European productions lose competitiveness, which could not be offset at the border, since information is not enough to operate in a non-discriminating way, and therefore in compliance with WTO regulations.

It must be also clear that the CAT would not require the immediate suppression of the ETS system, at least as long as its efficacy is not actually proven, since theoretically it could just be complementary and not alternative to *cap and trade*. It could also be in addition or complementary to the quantified emission reduction commitments of the EU Member Countries.

Unlike the ETS system, the CAT would also have the merit to afford a lower degree of arbitrariness (in determining the national emission ceilings) and to be easily re-adjustable according to the emission reduction targets.

Obviously, for the actual introduction of a CAT system, further scientific, technical and organizational aspects will have to be taken into consideration and they will require further studies and in-depth analyses, such as, for example:

- the study of the possible different methodology application options;
- in-depth analysis of taxation, legislation and international trade issues;
- evaluation, in terms of emission reduction, of the CAT efficacy on the economic system, also comparing different options with respect to different systems (e.g. ETS);
- evaluation of the CAT effects on consumers' behavior.

In order to regain credibility with the public opinion, the EU has to innovate their policy to combat global climate-altering gas emissions, currently so uncertain. The time has now come for proposals for new, innovative mechanisms, such as the above, to be included, with full rights, in the debate and in the analyses by those institutions and research organizations that are taking charge of finding a solution.

Relatedness, Economic Complexity and Convergence Across European Regions

Tullio Buccellato and Giancarlo Corò

Abstract The aim of this paper is to analyze how the heterogeneous structure of the European regions has affected their patterns of convergence or divergence. We analyse data collected by Eurostat, from a balanced panel of 191 regions and 55 economic branches over the period 2003–2015. In this way, we are able to describe and capture technological proximity across the regions and analyse how it has evolved over space and time. Limiting the analysis to the manufacturing activities, we are also able to measure the degree of economic complexity of the regional production systems and assess how this affects their patterns of growth.

Our findings suggest that spatial effects tend to push towards convergence, with the Eastern regions that started from relatively low levels of GDP per capita and experienced higher growth rates. Nevertheless, the different level of economic complexity tends to widen the gaps between territories: for example, the German regions, whose economic structures are more complex, have kept on widening the gap between themselves and the other European regions. The two different forces are also interconnected as the Eastern regions combine a relatively low level of GDP per capita with a significant level of economic complexity. During the period considered, the improvement in living standards has corresponded to the upgrade of their manufacturing production structures.

The current paper is an updated version of the working paper hosted on the University of Venice Ca' Foscari working paper series, here is the webpage link: https://iris.unive.it/retrieve/handle/10278/3717761/173862/WP_DSE_buccellato_coro_15_19.pdf

The opinions and views expressed in this paper by the author do not reflect in any way the opinions and views of Confindustria as an institution.

T. Buccellato (✉)
Economic Research Department, Confindustria, Rome, Italy
e-mail: t.buccellato@confindustria.it

G. Corò
Department of Economics, Ca' Foscari University, Venezia, Italy
e-mail: corog@unive.it

L. Paganetto (ed.), *Capitalism, Global Change and Sustainable Development*, Springer Proceedings in Business and Economics,
https://doi.org/10.1007/978-3-030-46143-0_10

Keywords Regional disparities · Growth · Convergence · Structural change · Relatedness · Economic complexity · Spatial effects.

1 Introduction

From the second half of the 1990s to the 2000s, the EU single market has undergone a period of fast political and economic integration, which experienced an important milestone in 2004, when ten new countries became Member States. At the same time of the EU integration, globalization accelerated worldwide, with a number of important economies, especially China, entering the World Trade Organization; the strengthening of links between countries has boosted trade, capital and migration flows. While it has been beneficial to many people all over the world, the enhanced degree of competition and the continuous speeding up of innovation have also enlarged the economic divide between nations and regions. Highly competitive territories have acquired a key advantage from the enhanced level of export, which, in turn, has brought higher returns on innovation, widening the gap between stronger and weaker regions, in terms of both knowledge and prosperity. As highlighted in recent reports (see Iammarino et al. 2018; Rodríguez-Pose 2017; Demertzis et al. 2019), in the last two decades, the tendency toward a geographical polarization has notably grown. Additionally, in the European Union, the inequality among regions has sharply increased, threatening social cohesion within and between countries and becoming one of the hardest political challenges for governments to deal with.

A key factor explaining the different paths of regional development is the uneven distribution of the productive knowledge, as well as the different learning skills acquired through the experience of the production (Stiglitz and Greenwald 2014). In this sense, the structure of economic systems well captures the set of competences available locally. In fact, convergence clusters often mirror groups of territories with related knowledge basis that tend to show both similar patterns of economic growth and a comparable evolution of the technological development (Boschma 2005). Therefore, the gaps in prosperity across territories are deeply influenced by their stock and the "quality" of the competences available locally; from this perspective, the "convergence clubs" mentioned by Baumol (1986) can be intended as clusters of countries with a similar knowledge basis and institutional setting.

The theory of economic complexity has offered an additional key to interpret the generation and persistence of gaps across nations and regions (Hidalgo et al. 2007; Hausmann and Hidalgo 2010; Tacchella et al. 2012; Hausmann et al. 2011), on the basis of the local stock of competences, in terms of its diversification, and their "quality", intended as their ubiquity (the more ubiquitous the less sophisticated). The theory of economic complexity focuses on productive knowledge embedded within real economic goods. However, "accumulating productive knowledge is difficult. For the most part, it is not available in books or on the Internet. It is embedded in brains and human networks. It is tacit and hard to transmit and acquire. It comes from years of experience more than from years of schooling" (Hausmann et al. 2011).

The structure of the economy situated in a given territory is therefore a key factor which explains both the patterns of economic growth and the evolution of the local knowledge. Accordingly, convergence and divergence across countries and regions has a twofold connotation, as it originates from the divide in productive competences and results in disparities in income and wealth. This reconciles the neoclassic perspectives of endogenous growth in terms of knowledge (Lucas 1988; Romer 1986, 1990) and learning by doing, with the patterns of technological change in the evolutionary economic theory (Nelson and Winter 1982).

The empirical analysis develops from descriptive statistics of economic development across the regions in the EU28 and discusses such evolution on the basis of the initial level of competences, captured by the structure in the economic branches. We then focus on the period 2003–2015 about which more detailed data are available, both on 191 regions and on 55 economic branches (i.e. comparing structural similarities as the parameter to measure the distance). Specifically, we construct an econometric model which takes into account both spatial and structural characteristics of EU regions to assess how these interactions influence convergence and divergence. Finally, we construct an index of economic complexity and show how it impacts the patterns of growth across the regions.

Our findings suggest that convergence processes in EU are driven mainly by a cluster of manufacturing regions belonging to Eastern European countries. Such regions took advantage of the set of manufacturing competences, developed during the communist era. Over that period, they underwent a process of quick modernization thanks to the fast integration with the value chain of the German and other productive markets. Such regions have been able to upgrade their competences and reach a more complex production process, which also has rapidly improved their level of GDP per capita in comparison to other mid-income regions in Western Europe. Divergence has been instead exacerbated by the dynamics of growth of the German regions, which, on average, have outperformed all the other regions, thanks to their highly complex manufacturing production fabrics.

In the next section we briefly introduce stylized facts on convergences and divergences between the European regions. The third section introduces the methodology used to analyze the structural similarity across the regions, it presents the econometric model used and illustrates our empirical findings. Section 5 concludes.

2 Stylized Facts on Convergence/Divergence Across European Regions

Looking at the patterns of prosperity across European regions over the last two decades, a general improvement in the living standards can be observed, although this has been accompanied by an increase in the absolute gap between richer and poorer regions. In 2000, the bottom group of regions (clustered in the first percentile level of GDP per capita at purchasing power standards, PGDP), had an average PGDP of 4200 euros; in 2015, the same statistical data increased by more than double the amount, to 9600 euros (Table 1). The growth pace of the PGDP for the regions at the bottom percentile of the distribution was faster in comparison to those at its top (the top/bottom ratio decreased from 10.3 to 6.2). Even if the absolute divide between the richer and the poorer regions has widened (the difference between the top percentile and the bottom percentile has increased from 39,200 to 49,900 euros), while the overall dispersion (measured by the relative standard deviation) remains stable. Furthermore, the distance between the regions at the top of the distribution and those around the median has been more persistent (the top/median ratio has remained stable at 2.3).

However, this overall stability hides very diversified paths of convergence and divergence among the regional economies. Figure 1 gives a quick overview of the change in ranking of the regions (see horizontal axis), compared to the initial level of PGDP (logarithm) in 2003 (see vertical axis). The graph can be sub-divided in four areas. First, the North-East quad, where regions have been climbing up the ranking, even if starting from better-off positions, i.e. exacerbating the divergence by contributing to the gap in the higher end. Second, the North-West quad, where regions have been losing positions in the ranking but started with a relatively high level of PGDP, i.e. fostering the convergence in the high end. Third, the South-West quad, where regions have been losing positions in the ranking, even if they started from relatively poor initial conditions in terms of PGDP, i.e. exacerbating the divergence in the low end. Finally, the South East quad comprises regions which have been gaining positions and started from relatively poor initial conditions, i.e. fostering the convergence in the low end. At a first glance, if any convergence took place over the period considered, this happened in the bottom part of the distribution, mainly due to the rapid growth in GDP per capita across regions in Central and Eastern Europe (see also Cuaresma et al. 2014).

Figure 2 shows the importance of Eastern European regions in driving the process of convergence. The red fitted line shows that an estimated rate of convergence of 3.1% (with a p-value below 1%) took place over the period 2003–2016; such results change substantially when the regions in Eastern Europe are not considered: the beta-coefficient get closer to 0 and loses significance. The regions which have contributed more to such result include: the region surrounding Bucarest in Romania, the Mazovia province and the lower Silesian province in Poland, where Warsaw is located, the Southwest Planning Region in Bulgaria, the region of Sophia, the East area of the Czech Republic.

Table 2 reports the top 20 of regions for the number of positions gained in the ranking of PGDP. The first factor that emerges clearly from the table is that geography considerably impacted the success of regions: apart from Malta, the top 20 are only by regions in Eastern Germany and Eastern Europe. Of particular interest in this sense is the analysis of groups of regions starting with a similar and relatively low level of GDP per capita but experiencing different paths of growth and development. For instance, Boltho et al. (2018) compare the cases of the regions in

Table 1 Summary statistics of the distribution of income in 2000 and in 2015

2000					2015				
	Percentiles	Smallest				Percentiles	Smallest		
1%	4200	3600			1%	9600	8400		
5%	6700	4200			5%	13,900	9500		
10%	8000	4200	Obs.	260	10%	15,600	9600	Obs.	269
25%	14,250	4200			25%	20,900	9900	.	
50%	19,200		Mean	19,225	50%	25,900		Mean	27,761
		Largest	Std. Dev.	9042			Largest	Std. Dev.	13,003
75%	23,300	36,000			75%	32,100	59,200		
90%	28,500	43,400			90%	39,400	59,500		
95%	32,300	48,700	Skewness	2.78	95%	46,700	76,200	Skewness	5.0157
99%	43,400	99,100	Kurtosis	25.36	99%	59,500	167,500	Kurtosis	51.61

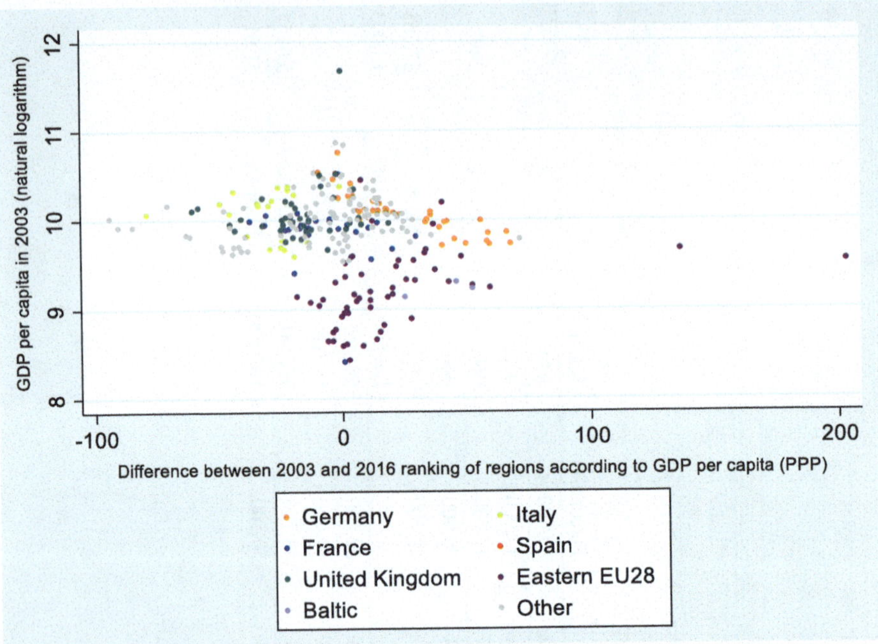

Fig. 1 The regions which contributed more to convergence 2003–2016

Eastern Germany and those of the Italian Mezzogiorno, concluding that the former were more successful in catching up with the rest of the country than the latter. This happens because of the higher homogeneity in terms of economic complexity, as national integration, per se, brings convergence in consumption rather than in GDP per capita.

The case of the reunification of Eastern with Western Germany is perhaps emblematic to explain the rapid catch-up of Eastern European regions with the remaining part of the EU. One would think that the reunification between the two German blocks has exemplified the integration of the Eastern economies in the Union. Germany has indeed played an important role, being far the largest foreign investor in the new countries, while Poland has been the most attractive place for German firms. As in the case of the Ost-Länder after the fall of the Berlin wall, Central and Eastern European regions have enjoyed a quick political integration, supported by the EU institutions. They also benefitted from a great amount of resources to modernize the infrastructure thanks to the EU policy of cohesion. Finally, the convergence of the private sector has reached highly competitive standards thanks to the foreign direct investments, especially by German corporations.

A clear example is the quick development of German corporations active in the automotive sector that, after the reunification, have expanded in the Eastern Ländern. For instance, the Eastern Region of Saxony, which hosted the obsolete automotive

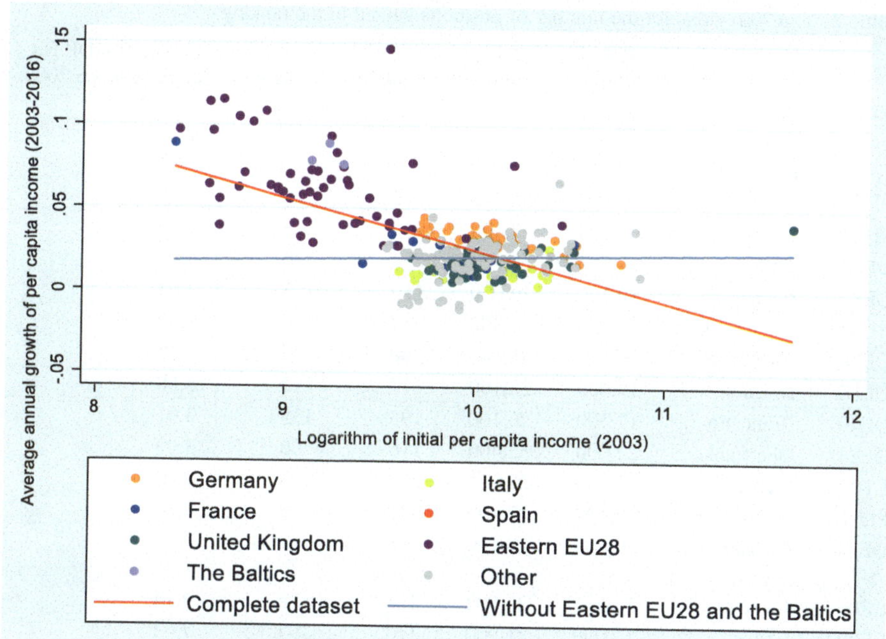

Fig. 2 Convergence across European regions over the period 2003–2016

industry inherited from the communist era at the beginning of the 1990s, by the early 2000s had already modernized the automotive cluster, serving mainly the Volkswagen corporation. A similar case study is the one of Berlin region, thanks to the heavy investments of BMW and Mercedes. Already during the 1990s, in parallel to the wide flows of investments in its Eastern Ländern, Germany had already started to redirect a great part of its foreign direct investments towards Eastern Europe, facilitating the entrance of such countries to the single market in 2004.

The contribution of our research is to investigate whether, in addition to the geographical factors, the initial structure of the regional economies might have played a role in determining their patterns of growth and development. Sharing the point of view of economic complexity, the productive structure of a territory mirrors the pool of knowledge, skills and know-how available at the local level, setting the basis for the learning processes from which long-term economic development originates.

Initial evidence is provided by the analysis of the structural change process for the top performing regions in comparison to the remaining one, over the period

Table 2 Top 20 regions for the number of positions gained in the ranking

NUTS2	Region	GDP per capita in 2003	GDP per capita in 2016	Ranking 2003	Ranking 2016	Growth of GDP per capita between 2003 and 2016 (%)
RO32	București—Ilfov	13,900	40,400	229	26	14.67
PL12	Mazowieckie	15,800	31,700	212	76	7.74
MT00	Malta	17,600	27,800	190	119	4.46
DEG0	Thüringen	16,900	26,700	198	130	4.46
DED5	Leipzig	19,200	29,000	170	103	3.93
BG41	Yugozapaden	10,300	22,800	244	184	9.34
DED4	Chemnitz	16,800	25,800	200	141	4.12
DED2	Dresden	19,000	28,100	172	113	3.68
DE40	Brandenburg	17,300	26,100	193	135	3.91
DE24	Oberfranken	22,000	33,400	116	60	3.99
LT00	Lietuva	10,200	22,000	245	193	8.90
PL51	Dolnoslaskie	10,600	22,100	242	190	8.35
DEE0	Sachsen-Anhalt	16,900	25,100	198	148	3.73
CZ06	Jihovýchod	14,600	23,600	224	176	4.74
EE00	Eesti	11,000	21,900	240	194	7.62
DE93	Lüneburg	17,800	25,700	188	143	3.41
PL41	Wielkopolskie	11,000	21,700	240	197	7.48
DE94	Weser-Ems	21,800	31,500	122	79	3.42
DEB1	Koblenz	21,600	31,100	125	84	3.38
SK01	Bratislavskj Kraj	26,900	53,700	47	6	7.66

2003–2015.[1] Figure 3 compares the difference in the productive structure in the initial and in the final year. Top performers had an initial weight of motor vehicles and machinery manufacturing, which was remarkably higher in 2003 and increased even more in 2015. This suggests that the initial competences in these two manufacturing branches have played an important role in the success of the top performing regions, which, in turn, have leveraged on the initial competences to further increase the size and scale of the activities in these branches.

We can also observe that the regions where the specialization in the tourism sector tends to prevail at the beginning of the period, the performance worsens over the period considered. In our opinion, this result is consistent with the assumption of economic complexity, according to which the learning processes are mostly nurtured by the variety of practical knowledge used in manufacturing (Buccellato 2016; Buciuni et al. 2014; Hidalgo 2015; Pisano and Shih 2012). Where the tourist supply

[1]To this purpose a mapping has been realized between the two different sector classifications used by Eurostat before and after 2007. Data are available over the full period for 11 regions out of the top 20 and for other 213 regions. Data have been reclassified in 55 economic branches.

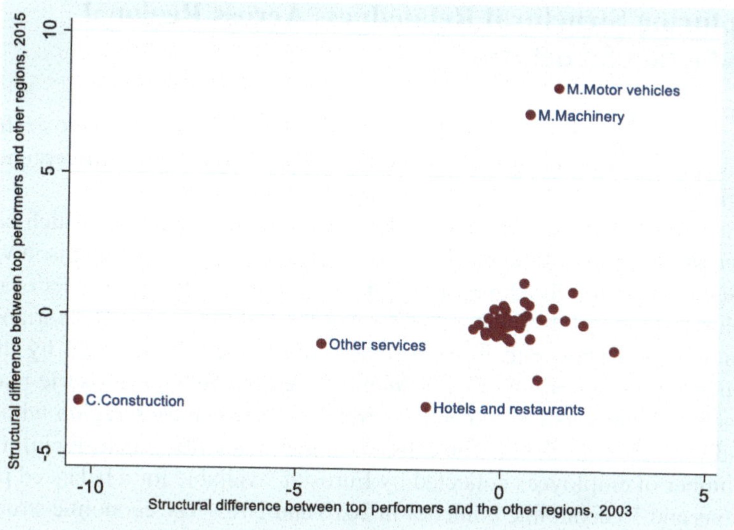

Fig. 3 Structural change in top performers with respect to other regions

dominates the economy, there is a tendency to present two effects that are not very helpful for long run growth: first, because the lacking of other industries, tourism demand feeds external production flows, with a low multiplier for the local economy; second, a dominant tourism economy risks to follow the logic of rent seeking, with reduced incentives for learning and innovation.

In general, the reasons of the persistent gaps across regions are multifaceted and find their roots in historical, institutional, technological and geographical factors. Since the 1990s, a large strand of the economic literature has been focusing on identifying the most meaningful determinants of convergence/divergence across countries and regions, following the path explored by Barro and Sala-i-Martin (1991) on the basis of the theoretical implications pointed out by the Solow (1956) and Swan (1956). A wide range of hypotheses on the key factors for development and growth have been tested with such stream of models, leading sometimes to contrasting results, depending on the theoretical specification selected—the so called "open-endedness" of growth theories (Brock and Durlauf 2001).

At the same time, the theory of economic complexity proposes an understanding of development based on the social ability to accumulate, share and create productive knowledge, which therefore tend to feed through learning-by-doing processes influenced by the pre-existing economic structure. The economic growth of regions will therefore be stronger if there is a diversified production base that enables the firms to employ and combine complex knowledge.

3 Capturing Structural Relatedness Across Regional Production Structures

In this section we propose a model to test how the structure of regions affects its future patterns of economic growth and the process of convergence/divergence with other regions.

The first step is the construction of the structural square matrix, which captures the degree of proximity of the product spaces of each region in comparison with the others. Such matrix has the dimension 191, corresponding to the number of regions for which the structural data are available over the full time span. The variable used is the amount of wages paid in each region and branch, normalized by the total amount per region. Based on this information, we construct a symmetric matrix, in which each cell contains the pair-wise correlation between each region with all the others individually considered. The cells of the matrix are filled up using information on the number of employees collected by Eurostat, available for a balanced panel of 191 regions and 55 economic branches in 2010 and 2015. The economic structure of each region is normalized by the total number of employees.

We are therefore able to construct a product space characterized by the technological proximity of regions. Such information is then used in a spatial econometric model, which has been found very useful for studying convergence across regions in various geographical contexts (Rey and Montouri 2000; Le Gallo et al. 2003; Arbia et al. 2005; Buccellato 2007). The originality of the model used in this specification is that we use the lag of the error term to capture relatedness across the regional production systems, in addition to the spatial lag of the dependent variable, to control for the effects of geographical distance across the observations. The model specification can be formally depicted as follows:

$$\frac{1}{T}\times\left(\frac{y_{i,t}}{y_{i,0}}\right)=\alpha+\beta\ln\left(y_{i,0}\right)+\rho W^S\left[\frac{1}{T}\times\left(\frac{y_{i,t}}{y_{i,0}}\right)\right]+\gamma X_i'+u_i$$

$$u_i=\lambda W^T u_i+\varepsilon_i$$

$$\varepsilon_i\tilde{i}.i.d.\left(0,\sigma^2 I_n\right)$$

where $y_{i,\,t}$ is the GDP per capita of region i as of date t, T is the length of the period, α is a constant and β is the convergence coefficient, the matrix X contains additional explanatory control variables and the respective vector of associated coefficients γ. W^S is the space matrix containing the inverse of the geographical distance and W^T the structural proximity of regional production structures. u_i is a non-spherical disturbance that is auto-correlated with respect to the degree of similarity of the industrial production fabrics.

As benchmark, Table 3 displays results related to the unconditional convergence for 268 regions for the whole period 2003–2016 and for the two sub-periods 2003–2007—before the economic crisis—and 2010–2016 –following the crisis

Table 3 Unconditional convergence with GLS

Variables	(1) Average annual growth in GDP per capita 2003–2016	(2) Average annual growth in GDP per capita 2003–2007	(3) Average annual growth in GDP per capita 2008–2016
Initial GDP per capita in logs	−0.0334*** (0.00327)	−0.0396*** (0.00532)	−0.0143*** (0.00242)
Constant	0.358*** (0.0326)	0.445*** (0.0531)	0.158*** (0.0244)
Observations	268	268	268
R-squared	0.382	0.302	0.106

Robust standard errors in parentheses
***$p < 0.01$

but excluding its two initial years. Convergence appears to be taking place over the full period considered (the beta coefficient is in the order of −0.0334 and significant at the 1% confidence level) with a more pronounced pace in the time span preceding the economic crisis (−0.0396), more than three times faster in comparison with what happened afterwards (−0.0116). The loss of momentum of the convergence process turns out to be evident also from the R-squared which drops from around 0.302 in the earlier period to 0.066 in the end.

Table 4 reports the same results introducing the full model presented above. The number of observations falls from 268 to 191, i.e. the number of regions for which structural data are available. The sample reduction does not seem to substantially affect the results regarding the convergence coefficients, which, even if with a slightly different magnitude, follow the same pattern over time, i.e. with a slower convergence pace after the economic crisis. Spatial effects are present over the full time span and with a stronger magnitude in the last period. The degree of similarity in the structure of the economies is also significant but loses significance when isolating the years preceding the crisis. These results suggest that the crisis has slow down the convergence processes across EU regions. However, after the crisis, the relatedness of the production structures started to affect more incisively the economic growth of regions, as it appears from the lambda coefficient modelled in the error term, which becomes very strong in magnitude and highly significant.

Figure 4 depicts the patterns of structural change across EU regions in terms of manufacturing activities, tending to be more complex as of the mix of competences involved in their production.[2] All the observations above the line tend to have experienced an increase in their manufacturing activities in 2015 in comparison with 2003. Among those areas which experienced a greater increase appear once again those in Eastern Europe—such result is more evident for the Romanian region of Vest and for the Hungarian regions of Közép-Dunántúl, Észak-Magyarország, Nyugat-Dunántúl. The regions consolidating their top positions are the key locations

[2]We have aggregated the manufacturing activities relating to the following sectors: chemical, pharmaceutical, electronic, electrical, machinery and motor vehicles.

Table 4 Unconditional convergence with GLS spatial regression including a spatial lag of the dependent variable and an error term auto-correlated according to the regional production structure estimated through 2SLS

Variables	(1) Average annual growth in GDP per capita 2003–2016	(2) Average annual growth in GDP per capita 2003–2007	(3) Average annual growth in GDP per capita 2010–2016	(4) Average annual growth in GDP per capita 2003–2016	(8) Average annual growth in GDP per capita 2003–2007	(12) Average annual growth in GDP per capita 2010–2016
Initial GDP per capita in logs	−0.0384*** (0.004)	−0.0423*** (0.007)	−0.0120*** (0.004)	−0.0331*** (0.003)	−0.0433*** (0.004)	−0.0126*** (0.003)
Spatial lag of dep.var.				0.824*** (0.109)	0.260** (0.131)	0.662*** (0.141)
Structural lag of in the error term				33.28*** (5.759)	−0.697 (1.649)	9.880*** (1.892)
Constant	0.408*** (0.0387)	0.472*** (0.0671)	0.144*** (0.0370)	0.334*** (0.0297)	0.470*** (0.0438)	0.136*** (0.0341)
Observations	191	191	191	191	191	191
R-squared	0.446	0.320	0.071			

Robust standard errors in parentheses
***$p < 0.01$

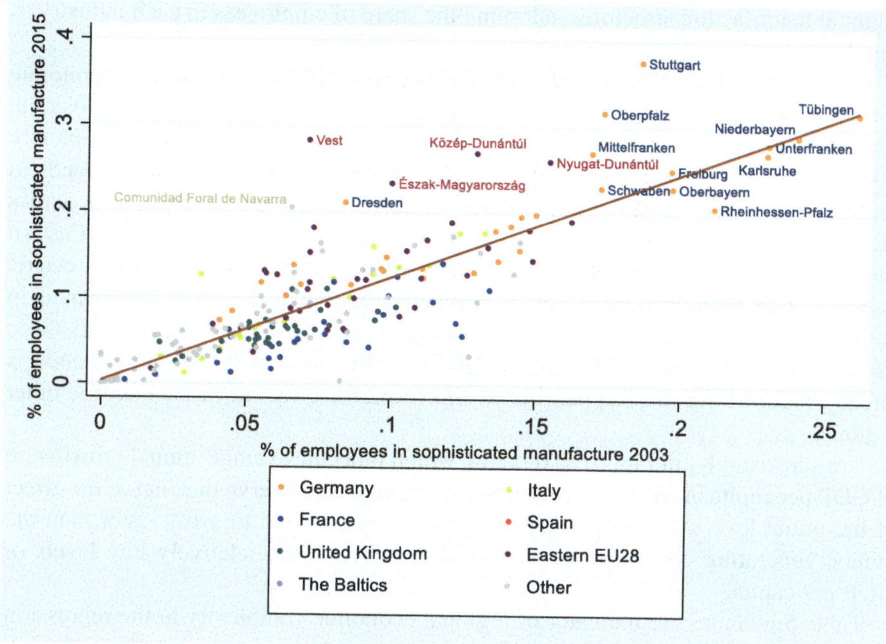

Fig. 4 The evolution in the structure of employees in relation to highly sophisticated sectors

of manufacturing activities in Germany—Stuttgart, Oberpfalz, Tübingen and Mittelfranken. Instead, French and British regions appeared to have undergone a period of reduction in the manufacturing activities, whereas the Italian regions were distributed close to the line, on average not having changed their structure significantly.

4 Economic Complexity and Regional Growth

The underlying assumption is that territories with comparable production structures display similar degrees of production knowledge and, hence, degree of economic complexity. To assess to what extent economic complexity might have affected the patterns of divergence and convergence across the EU regions, we use information on their production structure and on the average degree of product complexity in each sector, as shown in trade data. We compute a weighted average of product economic complexity, available from the Atlas of Economic Complexity, where the weight is provided by the value of trade of a given product category over the total value of trade, within the two digit HS product. This allows realizing a map between HS and NACE classification to compute the average product complexity of the regional manufacture structures, adopting the share of employees in each industry as weights.

Figure 5 (next page) shows the relation between GDP per capita and economic complexity in 2003 and in 2010. For countries, the relation tends to be positive. German regions are the ones exhibiting the higher degree of economic complexity associated with the higher level of GDP per capita. Economic complexity tends to capture very well how the divide in knowledge generates gaps in prosperity across the regions. What turns out to be evident, especially in 2003, is that the Eastern European regions tend to be all below the fitted line. This suggests that, ceteris paribus, such regions have a set of competences that is highly sophisticated in comparison with the income that they exhibit and this would imply a greater distance from their steady state. Therefore, higher growth rates. In 2010, the distance is already reduced but there appear to be still room for growing faster than the other regions.

Not surprisingly, in Fig. 6 (next page), which plots the average annual growth rate of GDP per capita in terms of economic complexity, we observe that, net of the effect of the initial level of GDP per capita, Eastern regions tend to grow faster than the others, generating convergence because they started from relatively low levels of GDP per capita.

Table 5 includes the indicator of regional economic complexity in the regression model to assess more precisely its impact on convergence patterns. Over the full period, it turns out to be strong in magnitude and highly significant. In the time span between 2003 and 2007, it remains positive but loses significance and, after the economic crisis, it exhibits stronger magnitude and impact. The convergence coefficient has a stronger magnitude when the one associated to the indicator of economic complexity loses importance. Such a pattern becomes even clearer when the spatial lag of the dependent variable is included. This evidence opens up the way to an additional interpretative hypothesis. In the first part of the sample, the Eastern regions have been catching up with the rest. Following the crisis, the convergence pattern has slow down, as the German regions have accelerated more than the others, generating again divergence.

5 Conclusions and Directions for Future Research

In this study we have analysed the patterns of convergence, in terms of structural change and GDP per capita across 191 regions part of the EU28. The analysis has been conducted adopting an original perspective of relatedness measured as the pairwise correlation between the distributions of wages paid in 55 economic branches in each region. This procedure has allowed constructing a spatial econometric model, encompassing also proximity in terms of similarity of the regional production fabrics. Based on such classification, we have studied the patterns of structural change in the network, over the period 2010–2016.

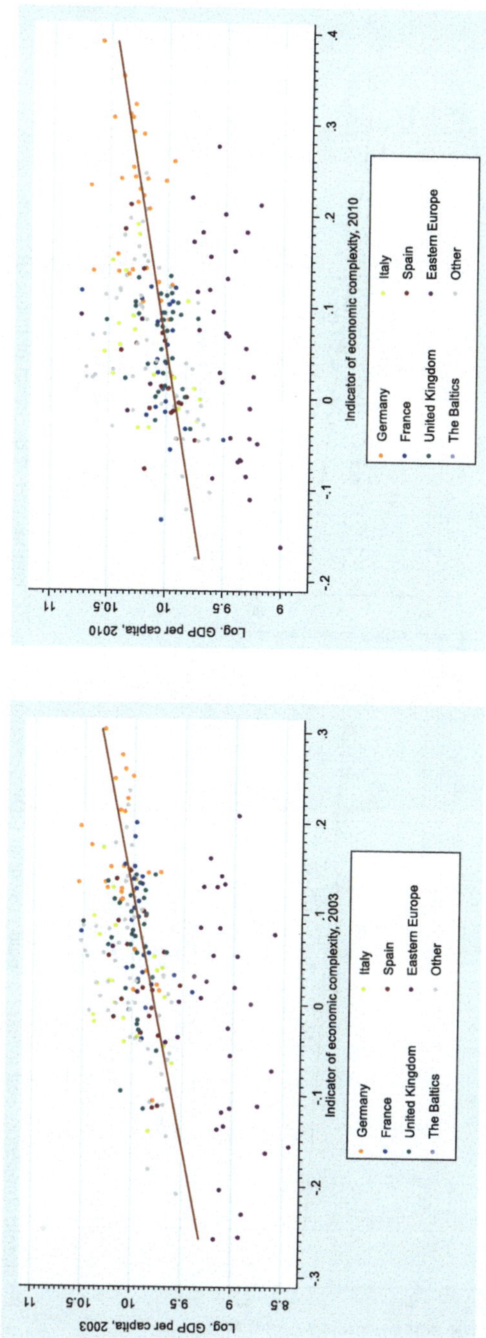

Fig. 5 Economic complexity and GDP per capita across EU regions

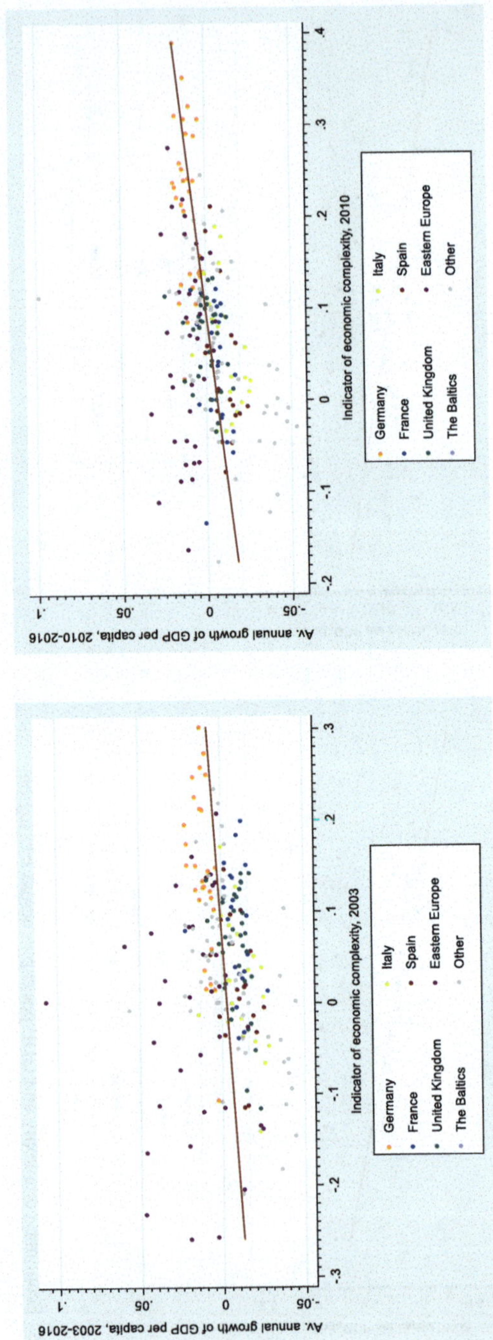

Fig. 6 Economic complexity and average annual growth in GDP per capita across EU regions (net of the effect of initial GDP per capita)

Table 5 Convergence and economic complexity, GLS and 2SLS

Variables	(1) Average annual growth in GDP per capita 2003–2016	(2) Average annual growth in GDP per capita 2003–2007	(3) Average annual growth in GDP per capita 2010–2016	(4) Average annual growth in GDP per capita 2003–2016	(5) Average annual growth in GDP per capita 2003–2007	(6) Average annual growth in GDP per capita 2010–2016
Initial GDP per capita in logs	−0.0426*** (0.00410)	−0.0444*** (0.00692)	−0.0185*** (0.00390)	−0.0382*** (0.00303)	−0.0441*** (0.00498)	−0.0190*** (0.00299)
Indicator of economic complexity	0.038*** (0.0134)	0.0195 (0.0202)	0.058*** (0.00884)	0.0146 (0.0127)	0.0105 (0.0212)	0.033*** (0.0113)
Spatial lag of dep.var.				0.884*** (0.0874)	0.251* (0.135)	0.677*** (0.135)
Constant	0.448*** (0.0404)	0.493*** (0.0688)	0.204*** (0.0386)	0.383*** (0.0300)	0.478*** (0.0493)	0.197*** (0.0297)
Observations	191	191	191	191	191	191
R-squared	0.465	0.323	0.192			

Robust standard errors in parentheses
***$p < 0.01$

Our results confirm that the patterns of convergence across EU regions are mainly driven by the rapid growth in terms of GDP per capita of Eastern European regions. In addition to this, it shows that such patterns have been accompanied to a marked shift towards more complex manufacturing activities. One possible explanation of this success story experienced in Central and Eastern European regions could be represented by the great participation of FDI especially originating from Germany (Crescenzi et al. 2017). Economic complexity of the production systems has helped the Eastern European regions to catch up in the early 2000s, but has also contributed to widen the gap between German regions and the rest in the years following the economic crisis of 2008.

Future research should investigate the drivers underlying the patterns of structural change further, in order to identify the triggers and obstacles for the upgrade of economic structures across territories. This should provide precious insights and bring better practices in areas that would otherwise result trapped into social and economic backwardness.

References

Arbia, G., Basile, R., & Piras, G. (2005). *Using spatial panel data in modelling regional growth and convergence*. Working Paper 55. ISAE.

Barro, R. J., & Sala-i-Martin, X. (1991). Convergence across states and regions. *Brookings Papers on Economic Activity, 22*(1), 107–182.

Baumol, W. J. (1986). Productivity growth, convergence and welfare: What the long run data show. *American Economic Review, 76*, 1072–1085.

Boltho, A., Carlin, W., & Scaramozzino, P. (2018). Why East Germany did not become a new Mezzogiorno. *Journal of Comparative Economics, 46*, 308–325.

Boschma, R. A. (2005). Proximity and innovation: A critical assessment. *Regional Studies, 39*, 61–74.

Brock, W., & Durlauf, S. (2001). Growth empirics and reality. *World Bank Economic Review, 15*, 229–272.

Buccellato, T. (2007). *Convergence across Russian regions: A spatial econometric approach*. Discussion Paper 70. Center for Financial and Management Studies, SOAS, University of London.

Buccellato, T. (2016). *The competences of firms are the backbone of economic complexity*. SSRN Working Paper.

Buciuni, G., Corò, G., & Micelli, S. (2014). Rethinking the role of manufacturing in global value chains: An international comparative study in the furniture industry. *Industrial and Corporate Change, 23*(4), 967–996.

Crescenzi, R., Datu, K., & Iammarino, S. (2017). European cities and Foreign investment networks. Scienze Regionali, Fascicolo 2, maggio-agosto 2017. https://doi.org/10.14650/86465

Cuaresma, C. J., Doppelhofer, G., & Feldkircher, M. (2014). The determinants of economic growth in European regions. *Regional Studies, 48*(1), 44–67. ISSN 1360-0591.

Demertzis, M., Sapir, A., & Wolff, G. (2019, April). Promoting sustainable and inclusive growth and convergence in the European Union. Bruegel Policy Contribution, Issue 7.

Hausmann, R., & Hidalgo, C. A. (2010, November). *Country diversification, product ubiquity, and economic divergence*. Faculty Research Working Paper Series, Harvard Kennedy School.

Hausmann, R., Hidalgo, C. A., Bustos, S., Coscia, M., Chung, S, Jimenez, J., Simoes, A., & Yıldırım, M. A. (2011). *The Atlas of economic complexity—Mapping path to prosperity*. Center

for International Development at Harvard University, Harvard Kennedy School, Macro Connection—MIT Media Lab, MIT.
Hidalgo, C. A. (2015). *Why information grows. The evolution of order, from atoms to economy.* New York: Basic Books.
Hidalgo, C. A., Klinger, B., Barabási, A.-L., & Hausmann, R. (2007). The product space conditions the development of nations. *Science, 317*, 482.
Iammarino, S., Rodriguez-Pose, A., & Storper, M. (2018). Regional inequality in Europe: Evidence, theory and policy implications. *Journal of Economic Geography*, 1–26. https://doi.org/10.1093/jeg/lby021.
Le Gallo, J., Ertur, C., & Baoumont, C. (2003). A spatial econometric analysis of convergence across European regions, 1980–1995. In B. Fingleton (Ed.), *European regional growth* (Advances in spatial sciences) (pp. 99–129). Berlin: Springer-Verlag.
Lucas, R. E. (1988). On the mechanics of economic development. *Journal of Monetary Economics, 22*, 3–42.
Nelson, R. R., & Winter, S. (1982). *An evolutionary theory of economic development.* Cambridge, MA: Harvard University Press.
Pisano, G. P., & Shih, W. C. (2012). *Producing prosperity. Why America needs a manufacturing renaissance.* Boston: Harvard Business School Press.
Rey, S. G., & Montouri, B. D. (2000). U.S. regional income convergence: A spatial econometric perspective. *Regional Studies, 33*, 143–156.
Rodríguez-Pose, A. (2017). The revenge of the places that don't matter. *Cambridge Journal of Regions, Economy and Society, 11*(1), 189–209. ISSN 1752-1378. https://doi.org/10.1093/cjres/rsx024.
Romer, P. M. (1986). Increasing returns and long-run growth. *The Journal of Political Economy, 94* (5), 1002–1037.
Romer, P. M. (1990). Endogenous technological change. *The Journal of Political Economy, 98*(5), s71–s102.
Solow, R. M. (1956). A contribution to the theory of economic growth. *Quarterly Journal of Economics, 70*(1), 65–94.
Stiglitz, J. E., & Greenwald, B. C. (2014). *Creating a learning society: A new approach to growth, development, and social Progress.* New York: Columbia University Press.
Swan, T. W. (1956). Economic growth and capital accumulation. *Economic Record, 32*, 334–361.
Tacchella, A., Cristelli, M., Caldarelli, G., Gabrielli, A., & Pietronero, L. (2012). *A new metrics for countries' fitness and products' complexity. Scientific Reports, 2*, 723.

Individual Behavior and Collective Action: The Path to Iceland's Financial Collapse

Thorvaldur Gylfason and Gylfi Zoega

Abstract Unsustainable accumulation of debt precedes financial crises. The recent Western financial crisis was no exception in this regard. The external debt of Greece, Iceland, Ireland, and Spain increased exponentially, in Iceland at a rate higher than the rate of interest on foreign debt. The Ponzi scheme that played out in Iceland begs the question why a country would set out on a path that could lead to a financial crisis. We address this question and describe the private incentives faced by bankers, financiers, politicians and others. In particular, we show how private incentives and a culture that valued financial gains above all else collided with socially desirable outcomes. The root of the problem in Iceland as well as in other crisis countries was a failure at the state level to align private incentives with what was socially prudent, a failure due, at least in Iceland, to a combination of mistakes, incompetence and what can only be called corruption. Furthermore, misplaced belief in a market economy where morals and ethics play no role paved the way to serious lapses in accounting and in the operation of the banks.

Keywords Financial crises · Corruption · Culture · Iceland · Quality of governance · Rent seeking

---

The authors are indebted to Gunnar T. Andersen, Lars Jonung, Gylfi Magnússon, and Stefán Svavarsson for their helpful reactions to an earlier version of the text. A preliminary version of this paper appeared as CESifo Working Paper 7874, 2019.

T. Gylfason (✉)
University of Iceland, Reykjavík, Iceland

CESifo, Munich, Germany
e-mail: gylfason@hi.is

G. Zoega
University of Iceland, Reykjavík, Iceland

CESifo, Munich, Germany

Birkbeck College, London, UK
e-mail: gz@hi.is

L. Paganetto (ed.), *Capitalism, Global Change and Sustainable Development*, Springer Proceedings in Business and Economics,
https://doi.org/10.1007/978-3-030-46143-0_11

JEL E44 · G01 · G41

1 Introduction

The international monetary regime that emerged from the collapsed Bretton Woods system in the early 1970s has proved unstable. Several waves of financial crises have occurred in the past four decades. There was the crisis in Mexico and South American countries in the early 1980s; then Japan, the Nordic countries (except Denmark) and Mexico in the early 1990s; the crisis in South East Asia in 1997; Russia in 1998; and the Western financial crisis in 2008.[1] Each of these episodes followed a similar pattern. Large current account surpluses in other countries released capital inflows into recipient countries where credit expanded, currencies appreciated (in countries with floating rates) and asset prices increased. Higher real estate and equity prices and elevated exchange rates then increased consumption and investment causing current account deficits. A sudden stop of the capital inflows then made asset prices and the exchange rate collapse, triggering a banking crisis, a currency crisis and in some cases a sovereign debt crisis.

While the mechanisms of the boom and bust seem clear, what remains to explore is why some countries and not others welcomed the destabilizing capital inflows. This is our aim here. We use Iceland, our native country, to decipher the domestic causes of the inflows. We ask why an affluent country like Iceland, or rather its bankers, egged on by business leaders and politicians, chose to borrow abroad at an unsustainable rate.

History shows that unsustainable accumulation of debt, private or public, precedes financial crises (Reinhart and Rogoff 2009). We use the experience of Iceland's financial collapse in 2008 to explore why borrowers and lenders engaged in such a scheme that was bound to collapse as, among others, senior Central Bank officials had realized already in 2006 and they admitted under oath before a specially convened Court of Impeachment in 2012. The starting point of our story is the potential conflict between private incentives and socially desirable behavior—what Olson (1971, 2000) called private and social rationality—as well as between productive and unproductive (i.e., rent seeking) activities (Krueger 1974). Further, we will explore how a political culture that encouraged a selfish quest for wealth—or greed, if you prefer—regardless of the external effects on others facilitated the boom and bust. Finally, we will argue that the government, the central bank and the financial supervisory authority failed to align private incentives with financial stability. Thus, the Iceland experience teaches us that in an unstable international financial system where destabilizing capital flows have caused one crisis after another, the countries that suffer the crises are the ones where the collective action needed to ensure financial stability fails.

[1] See Aliber (2016) and introduction in Aliber and Zoega (2011).

We begin by briefly reviewing Iceland's economic history from 1900 to date. In 1904, when Iceland attained home rule from Denmark, Iceland's per capita national income was about half that of the mother country. Thereafter, Iceland caught up, gradually attaining living standards broadly comparable with the rest of the Nordic region. It was a bumpy ride. We aim to illuminate, *inter alia*, Iceland's domination first by farming interests and later by its fishing industry until an attempt was made to diversify the economy by quickly turning Iceland into a global financial center in the late 1990s, resulting in the dramatic financial collapse of 2008.

Against this background, we tell the story of the events leading to the collapse of 2008, including the privatization of the banks and the institutional setup that allowed bank owners and their favorite customers to profit by putting others at risk. We then briefly recount the story of the crash and its immediate aftermath. Iceland's collapse was among the greatest financial crashes on record (Laeven and Valencia 2013), calling for the first IMF developed-country rescue operation in a generation. The bankruptcy of the three Icelandic banks combined was, in dollar terms, the third largest corporate bankruptcy on record after Lehman Brothers and Washington Mutual. The financial losses inflicted on foreign and domestic residents amounted to six times Iceland's annual GDP (International Monetary Fund 2009).[2] The malfeasance was considerable. By the end of 2017, the Supreme Court of Iceland had sentenced 36 bankers and others to a total of 88 years in prison for crash-related offenses, which is also unique (Jensdóttir 2017). Thus, although the economy has recovered there is a lingering mistrust toward public institutions and politicians.

We will discuss the social psychology of the aftermath of the crash, including the refusal of those identified by the parliament's Special Investigation Commission in 2010 as being primarily responsible for the crash to admit to mistakes. While the SIC described wrongdoing and mistakes by bankers, the central bank and the financial supervisory authority it did not scratch below the surface to explore the root causes of the behavior exposed by the crash.

The mentality and institutions involved in the Iceland case are deep-rooted in the country's political and cultural environment that is marred by structural flaws identified by the Constituent Assembly elected by the nation to draw up a new post-crash constitution for Iceland in 2011 (Gylfason 2013). These include the clan-based stratification of society and oligarchic nature of Iceland's natural resource management system as well as unequal suffrage—i.e., unequal apportionment of seats in parliament—that seem to foster pockets of social inefficiency and to allow the few to profit at the expense of the many.

[2]The IMF's initial estimate of seven times GDP can be reduced to six times GDP in view of better-than-expected asset recovery. See Gylfason (2015, 2019) and Benediktsdóttir et al. (2017). These estimates are incomplete, however, as they include, e.g., the decline in the value of stock-market and pension-fund assets from an artificially inflated value before the crash. On the other hand, they do not include the loss of net worth of the 10,000 households, one household in 12, that lost their homes. Homes lost to banks should count as losses rather than as cost-neutral transfers of wealth from households to banks. Several such complications arise in a comprehensive accounting of the total cost of financial crashes, which is beyond the scope of this study.

2 From History to Theory

Among the poorest countries of Europe in 1904 when Iceland was granted home rule after more than 600 stagnant years of belonging first to the Norwegian and then Danish crown, Iceland took off with fanfare as the twentieth century began. From 1900 to 2018, real per capita GDP increased by a factor of 20. This is not a misprint. Real per capita GDP grew by 2.6% per year on average during this long period compared with 1.6% per year in the United States 1776–2016 and 1.4% per year in Italy 1861–2016 (Maddison 2019). Remarkably, growth in Iceland was virtually the same during the first half of the period 1900–1959 as during the second half 1960–2018. Even so, output was volatile. The economy took several deep dives, for example when fish catches failed in the 1960s and when inflation was brought down in the 1980s. None of those dives was deeper than the one triggered by the collapse of the financial system in 2008. By 2010, per capita output was 10% less than in 2007, and was not restored to its 2007 level until 2015 or 2016. Hence, the IMF-assisted economic recovery from the crash took 8–9 years, the average length of recoveries from financial crises reported by Reinhart and Rogoff (2014).

It took many right decisions to catapult Iceland from misery to modernity in such a short time. The population as well as income per capita remained stagnant for centuries until the 1890s when the imports of sailing boats and later motorboats boosted fishing. The main drivers of growth were imported technology and capital that allowed the nation to use natural resources more extensively. Thus, growth was largely driven by improved inputs into production rather than by innovation. The education of the labor force improved greatly, aided by general adult literacy since the mid-1700s as well as by the influx of equipment, skills, knowledge, ideas, and attitudes that followed from hosting first British and then American troops during and after World War II. An increasingly mechanized fishing industry became Iceland's chief earner of foreign exchange and was able to exploit the fishing grounds within Iceland's economic jurisdiction that was gradually extended from three nautical miles from shore to 200, equal to 230 miles or 370 km. Hydropower and geothermal energy sources where harnessed from the 1960s onward for local use as well as for export through aluminum and ferrosilicon, lightweight energy-intensive products that are relatively easy to transport. Iceland liberalized its external trade regime by joining the European Free Trade Association (EFTA) in 1970 and the European Economic Area (EEA) in 1994, making Iceland a de facto 70% member of the EU. After 2008, tourism suddenly became Iceland's chief foreign exchange earner, larger than either the fisheries sector or the energy sector. If well managed, a great earning potential shared by a small population is conducive to high incomes per person, at least for a time, but tourism may prove to be a mixed blessing as argued by Ghalia and Fidrmuc (2015).

Before turning to the privatization of the banking system and the failed attempt to develop an international banking sector we first discuss the prevalence of rent, lack of competition, and the associated corruption in the recent economic history of Iceland, factors that have impeded growth as we will explain.

2.1 Economic Rent

Natural resources in the form of fish stocks and geothermal and hydro energy are a source of significant economic rent. The strategic location of Iceland has also generated rent in the form of transfers from the United States intended to ensure the goodwill of the population and acceptance of American forces on the island.[3] This rent accrued to private companies that were connected to the governing political parties.

The privatization of the banking system 1998–2003 generated economic rent through a high credit rating of the newly privatized banks due to their systemic importance. Thus, three large commercial banks benefited from being domiciled in a country where the sovereign had not defaulted in the past. This allowed the banks to borrow from foreign banks, which thought they knew that in case of default, there was a government with low levels of debt standing behind the borrowers and behind it all stood the International Monetary Fund. The consequences soon materialized in impressive acquisitions by the banks and their owners of foreign as well as Icelandic businesses. One more national resource, the sovereign's credit rating, had thus been privatized.

The history of rent seeking is a long one in Iceland and has absorbed the time and talent of many promising young people over the years. Shortly after the economy took off around 1900, the First World War and then the Second World War generated significant economic rent. Import restrictions were introduced to protect farmers during World War I. Having increased steadily from 25% in 1870 to 60% in 1915, the ratio of exports to GDP then decreased to 20% in 1945 (Fig. 1).[4] The period 1930–1960 saw pervasive trade restrictions extended across the board, leading to rampant rent seeking with economic distortions to match as described by Krueger (1974). During World War II, the arrival of U.S. forces in Iceland in 1941 followed by membership in NATO in 1949 also became a source of corrosive rent-seeking behavior as individuals with political party connections profited from the NATO base that remained open until 2006, adding about 2% to Iceland's GDP per year on average.[5] No estimates of the size of the rents involved are available. An Icelandic shipping company was given a monopoly on shipping goods from the United States to the military base in Iceland; a local construction firm was given a monopoly on all construction at the base; and a precursor to the current airline Icelandair was given the right to offer cheap flights across the Atlantic in the 1950s

---

[3]The U.S. government unilaterally withdrew its forces from Iceland in 2006 against the protests of the Icelandic government, protests best understood in the light of the reduction in rent accruing to the local economy.

[4]More than a hundred years later, import restrictions against farm products remain largely intact as high tariffs replaced an outright ban against importation of dairy products and meat.

[5]The contribution of the NATO base to the Icelandic economy was particularly large in the early years, amounting to 15%-20% of total foreign exchange earnings during 1953–1955, for example (Ingimundarson 1996, p. 282).

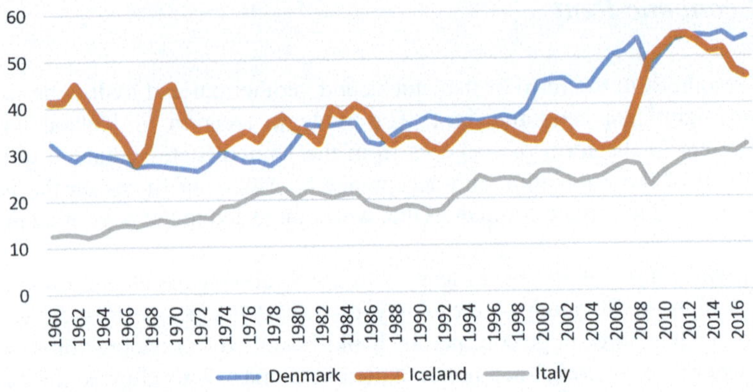

Fig. 1 Exports of goods and services 1960–2017 (% of GDP) (Source: World Bank, *World Development Indicators*)

and 1960s, long before President Carter deregulated the airline industry in 1978, plus a slot at Kennedy airport. In the 1950s, despite not having suffered any war damages apart from merchant ships that were sunk by German submarines, Iceland received Marshall Aid that exceeded on a per capita basis what all other European countries received. Further, the bulk of fish exports went to the U.S. destined for public institutions such as prisons and schools. This was economic rent created by the strategic location of the island, the "unsinkable aircraft carrier" as described Joseph Luns, Secretary General of NATO 1971–1984.

Estimates are available for more recent natural resource rents. The rent arising from Iceland's fisheries is currently estimated at 2–3% of GDP (Thorláksson 2015). This estimate has been roughly unchanged since the introduction of the catch quota system in the fisheries in the mid-1980s. Of this yearly amount, at present, about 10% accrues to the public, by law the rightful owner of the marine resources in Icelandic waters, through nominal fishing fees that were introduced in 2002. The rest accrues to vessel owners who, despite their small number, have emerged as an influential—and, unsurprisingly, reform-resistant—player in business and politics. It is plausible that a part of the rent accumulates in foreign banks since the larger vessel owners sell the catch to their own fish processing plants, which then sell the final product to foreign subsidiaries that sell it to the consumer. This invites double pricing, which in principle gives the businesses a possibility to let the rent appear abroad, perhaps in a low-tax entity.

The rent arising from Iceland's energy sector is also substantial, equivalent to around 1.5–2% of GDP (Jóhannesson 2015). Unlike in the fishing industry, the energy rent has not given rise to a class of local oligarchs. Instead, politicians sometimes promise aluminum smelters to their voters before elections, thereby weakening their negotiating position vis-à-vis the buyers of the energy for the partly foreign-owned smelters. This pattern of behavior apparently led for a long time to such low prices for the energy sold that the prices had to be kept secret from the owner of the energy, the general public. Further, the international aluminum

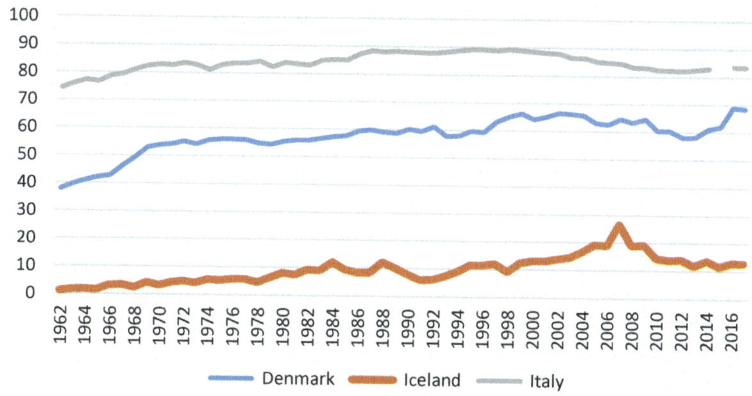

Fig. 2 Manufactures exports 1962–2017 (% of total exports) (Source: World Bank, *World Development Indicators*)

companies have managed to avoid taxes by transferring their profits to low-tax countries, again through companies based in low-tax countries, which then lend money to the operator of the smelters in Iceland, the interest on the loans surprisingly close to the profits from the smelters.

The combined stock value of Iceland's fish and energy resources is considered to amount to somewhere between 67 and 90% of GDP which, based on the middle value, is equivalent to roughly USD 170,000 for each family of four in Iceland (Jóhannesson 2015).[6] As public awareness of these issues and magnitudes increases (official statistical reports do not include this information),[7] public support for a more equitable disposal of the resource rents may increase. In a national referendum held by parliament in 2012 on a new post-crash constitution for Iceland, 83% of the voters declared their support for a constitutional provision stipulating national ownership of natural resources not in private possession (Gylfason 2013). Attempts remain to be made to assess the implicit monetary value of Iceland's natural environment, a major magnet for tourists.

The orientation of Icelandic exports toward natural resources and the associated systemic overvaluation of the ISK, a common manifestation of the Dutch disease, helps to explain why Icelandic manufactures account for only 13% of merchandise exports in Iceland compared with 69% in Denmark (Fig. 2). Iceland still produces few goods that other countries want to buy. Even so, significant diversification of total exports has taken place in that the fisheries are no longer the dominant export industry because tourism has become more important for foreign exchange earnings. This seems likely to reduce the political influence of the fishing lobby. A risk stems, however, from the fact that tourists can be fickle.

---

[6]For comparison, the Norwegian Pension Fund, earlier Oil Fund, the world's largest sovereign wealth fund, amounts to about USD 800,000 for each family of four in Norway.

[7]Iceland's fisheries rent is not included in the World Bank's recent tabulation of natural resource rents around the world from 1970 onward in the *World Development Indicators*.

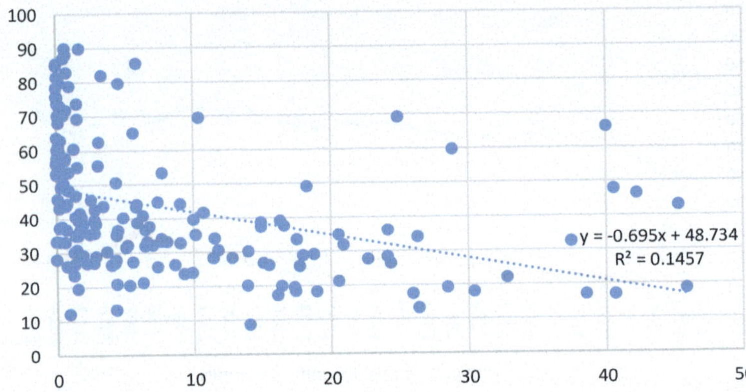

Fig. 3 Corruption 2012–2018 and Resource Rents 1970–2017 (Source: Authors´ computations based on data from World Bank, *World Development Indicators*, and Transparency International. Note: Vertical axis shows corruption perceptions index. Horizontal axis shows total natural resource rents in % of GDP)

Of the more than 200 countries for which the World Bank presents estimates of total natural resource rents, 95 countries have a higher share of rents in GDP than Iceland's 4– 5%. The world average is 2%. Figure 3 shows the cross-country relationship between the average share of natural resource rents in GDP 1970–2017 as reported by the World Bank and the 2012–2018 average of Transparency International's Corruption Perceptions Index that extends from 100 (clean) to zero (corrupt). The figure covers 177 countries. The correlation between the two variables is −0.38 and statistically significant (t = −5.5). Taken at face value, the slope of the regression line, −0.7, in Fig. 3 suggests that a drop in the natural resource rent share in GDP by 20 points, corresponding to the difference between, say, Angola at 29 and Ghana at 9, would in the average country go along with a 14-point reduction of the corruption index, spanning one sixth of the scale that reaches from ten in Somalia to 91 in Denmark.[8]

The orientation of the Icelandic economy toward natural resources spread learned rent-seeking behavior to other areas, including banking. After the privatization of the banks in 1998–2003, their new owners engaged in reputation mining that has many things in common with other forms of resource depletion. Sudden inflows of foreign credit exerted a similar manna-from-heaven effect on its recipients as resource windfalls (Gylfason and Zoega 2018). The source was the same in both cases. By allocating valuable common-property fishing licenses to select vessel owners first for free from 1985 onward and then for a nominal fee after 2002, parliament created a class of wealthy oligarchs. By delivering two of the three state banks to political

[8]Statistical endogeneity bias is hardly an issue here because corruption 2012–2018 cannot have exerted but a minor effect on resource rents 1970–2017.

friends during 1998–2003, parliament added to the earlier class of oligarchs (Exhibit A).

Exhibit A Privatization Among Friend
The privatization of Iceland's commercial banks 1998–2003 did not begin in earnest until after the bulk of the banking system in East and Central Europe had been privatized. Mid-stream, the government abandoned its original plan for dispersed ownership, including foreign owners, and decided rather to sell the banks at modest prices to local political allies. The buyers of two of the three banks lent each other a substantial part of the down-payment on their purchases. Further, the buyers of Kaupthing falsely claimed to have a foreign partner, a small German bank, to sweeten their tender (Júlíusson 2018). The collapse of all three banks a few years later resulted in total collateral damage—which includes both defaults on external debt as well as the collapse of equity value—equivalent to six times Iceland's GDP (see fn. 4), two thirds of which were inflicted on foreign creditors, depositors and shareholders and the rest on domestic residents as the local stock market was virtually wiped out overnight, pension funds whose managers had purchased bank stock as if there was no tomorrow took a big hit, and more. The defaults on private external debt amounted to 400% of GDP. The sometimes-huge write-offs of nonperforming loans following the crash are not public information case by case, raising concerns about possible discrimination among borrowers.

2.2 *Limited Competition*

The tardy trade growth described in Figs. 1 and 2 contributed to an associated weakness in the shape of excessive concentration of economic activity, including insufficient export diversification. True, specialization in production for export in keeping with comparative advantage generates gains from trade. Even so, excessive specialization can increase macroeconomic risk and volatility and thus destabilize or undermine economic growth, especially if the specialized sector becomes so dominant as to be able to damage other industries through rent seeking, repeated bouts of the Dutch disease, or distortions of the political system (Gylfason and Wijkman 2016).

The share of fish products in Iceland's total exports, at over 90% in the 1950s, declined below 50% in the 1970s and 1980s after the government, facing the need to regulate the fisheries to conserve marine stocks, decided to launch energy-intensive aluminum and ferrosilicon exports to supplement one natural-resource-based industry by another, a heterodox diversification strategy that, in effect, persists to this day. The sole significant deviation from this natural-resource-based strategy was the government-sponsored attempt to turn Iceland into an international banking center following the privatization of the banks 1998–2003, an attempt that daring bankers

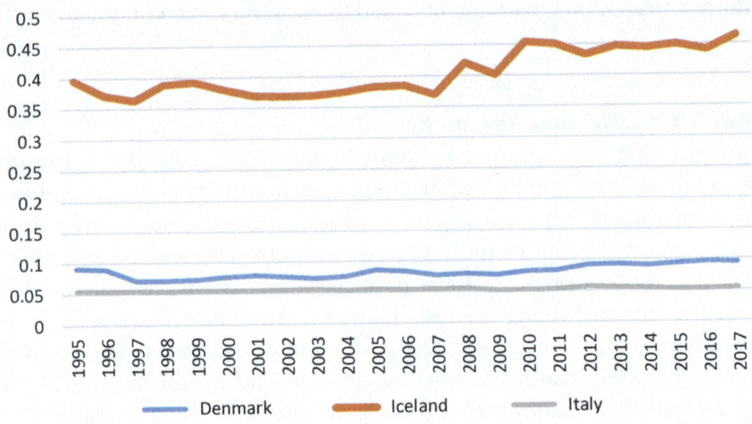

Fig. 4 Export Concentration 1995–2017 (Source: UNCTAD. Note: Vertical axis shows Herfindahl-Hirschman index. Rising curves mean more concentration)

apparently sold to—unless bought from is a better way to describe what happened[9]—receptive politicians and that ended in disaster with the collapse of the entire banking system in 2008. By contrast, the sudden rise of the tourism industry after 2008 was market-induced, helped by the depreciation of the króna that for a long time had been overvalued for reasons having to do especially with fisheries policy and high inflation (15% per year on average during 1960–2017), rendering Iceland too expensive in the eyes of foreign tourists and Icelandic tourism thus uncompetitive. During 2016–2018, tourism generated more foreign exchange than the fisheries and energy sectors combined.[10]

The concentration of Icelandic exports is borne out by UNCTAD statistics. The Herfindahl-Hirschman index (HHI) of market concentration, which ranges from zero (no concentration) to one (extreme concentration), is high in Iceland (Fig. 4). This is a country-specific index that is unrelated to market concentration in other countries and is defined as the sum of the squares of the shares of each sector of production in total output (or sometimes as the square root of the sum of squares). It covers only merchandise exports, i.e., exports of goods, not services. As a country's markets become more concentrated—i.e., less dispersed or, if you prefer, less diversified—the normalized value of the HHI rises toward one.[11]

[9]The Special Investigation Commission (2010, vol. 8, pp. 164–169) reports that during 2004–2008 the banks granted financial support to political parties and politicians in the amount, in today's money, of almost 20 euro per vote cast in the 2007 parliamentary election.

[10]The tourism boom in Iceland may not last, however. Ghalia and Fidrmuc (2015) study the relationship between tourism and growth in 133 countries during 1995–2007 and find that a dependence on both trade and tourism tends to reduce economic growth, an effect they attribute to the Dutch disease.

[11]The Finger-Kreinin index (FKI) of export diversification, a relative index that compares the structure of exports across countries by showing the extent to which the structure of exports by

The high concentration of Icelandic exports shown goes along with a lack of competition in local markets where oligopoly reigns supreme. The banking sector, which has always been and remains mired in controversy or worse, is still dominated by three banks with a combined market share of 97%. They operate in a seller's market, which enables them to discriminate among their customers, squeezing some while letting others off the hook. It was this power to discriminate that induced the government to privatize the banks 1998–2003 by delivering them to their cronies. Within Europe, Iceland is unique in that its banks face no foreign competition at home.

Likewise, a couple of large firms dominate the retail sale of petroleum with close albeit shifting connections to the two largest political parties, both now split and in deep decline since the financial crash. Thinking that business as usual was still permissible after Iceland joined the EEA, the two main oil companies that resell imported oil to local customers were found guilty of illegal collusion 1993–2001 that had inflicted significant costs on the public. The firms were fined and their CEOs were indicted but the case against them was ultimately dismissed. The combined market share of the three largest oil companies still exceeds 90%. For the first time, however, they will soon be exposed to foreign competition, an inconvenience that the banking system has not yet had to face.

The combined market share of the three largest insurance companies is also about 90%. They were all found guilty of illegal collusion 2002–2005. For a long time, two firms shared about 90% of the market for building materials. They were also found guilty of illegal collusion, and now face a foreign competitor at home. The pattern is clear.

Oligopoly is not an inevitable consequence of Iceland's small size. Just as small countries use foreign trade to compensate for their inability to produce many goods and services that only larger countries can produce, commodities that small countries need to import and pay for by their export earnings, they can also use foreign competition to protect domestic consumers against oligopolistic tendencies among local producers (Alesina and Spolaore 2003; Gylfason 2009).

---

product of a given country differs from the world average, tells the same story (not shown here). Both indices, the HHI and the FKI, vary inversely with country size as measured by output or population, because small size encourages specialization. When the HHI is adjusted for country size by taking as a measure of concentration the difference between the actual average HHI during 1994–2014 and the HHI predicted by a linear cross-country regression of HHI on the log of the average population during the same period, similar results obtain. But size does matter: Iceland moves from 76th place down to 98th place in a sample of 202 countries ranked by export concentration, where Iraq, Nigeria, and Chad head both lists. With the adjustment for population, Italy moves up the list from 202nd place to 187th. The adjustment method is taken from Gylfason (1999), where it was used to extract an index of openness to trade from exports-to-GDP ratios that, like the HHI and the FKI, vary inversely with country size as measured by population.

2.3 Corruption

The aim of the foregoing list of topics—trade restrictions, rent seeking and lack of competition—is to suggest why Iceland has been prone to corruption in the form of rent seeking as has recently begun to show up in international data (Gallup 2013; Transparency International 2018). The corruption takes the form of the political allocation of rent to private parties and the reciprocal relationship among political parties and large firms that receive the rent, including protection from competition. In effect, these companies are more powerful than others that only engage in productive activities and they have an incentive to protect their position by influencing politics and owning newspapers and other media outlets.

Krueger (1974) laid out the relationship between trade restrictions and corruption while Ades and di Tella (1999) estimated the relationship between competition, rents and corruption, showing how sheltering firms from foreign competition is conducive to corruption, especially in economies with a small number of firms. Their descriptions fit Iceland well as does the analysis of Pendergast et al. (2011) linking natural resource rent to corruption. Some other explanations of corruption on offer seem less relevant to Iceland such as Shleifer and Vishny's (1993) hypothesis that weak governments are conducive to corruption because they fail to keep corruption under control. The commonly held view that weak laws are to blame for corruption does not seem to apply to Iceland mainly because Icelandic law is quite like Danish law and European law through membership in the European Economic Area. Even so, a weak regulatory framework contributed to the recent financial boom and bust, as we will discuss below. Weak law enforcement may play a role. Another commonly held view is that low pay of politicians and public officials is a source of corruption (Van Rijckeghem and Weder 2001). This description does not seem to fit Iceland either because, at present at least, many Icelandic politicians and bureaucrats are better paid than in neighboring countries but then their high pay despite poor performance may itself signal corruption.[12] Gallup (2018) reports that only 18% of the Icelandic electorate trust parliament (Exhibit B).

Banking has long been a special source of corruption in Iceland because politicians could use the banks to further their economic and political interests. Already in their infancy in the first years of home rule in the early 1900s, the banks were mired in controversy and sometimes scandal. Politicians used them both as places of employment for themselves and their attendants and as sources of favorable loans to friends and allies, a practice that persisted throughout the twentieth century. Negative real interest rates for decades on end made the banks especially attractive as political instruments. Privatization of the banks during 1998–2003, even later than in most of East and Central Europe, was originally meant to sever the links between politics and banking but, in mid-stream, it was considered safer to preserve the

[12]For surveys of corruption and the corruption literature, see Ades and di Tella (1997), Tanzi (1998), Treisman (2000), Jain (2001), Aidt (2003), Rose-Ackerman (2006), and Rose-Ackerman and Søreide (2011).

umbilical cord connecting the banks to politicians. This proved to be a fatal error to which we now turn.

Exhibit B From Panama with Love
In 2012, humbled by the crisis, Parliament resolved unanimously that "criticism of its political culture must be taken seriously." [Our translation.] Then, in 2016, it came to light that the names of about 600 Icelanders, including three cabinet ministers, who were also Members of Parliament, were among those exposed in the Panama Papers (2016). Of the 332 cabinet ministers in Western Europe, five showed up in the Panama Papers and three of those five were Icelanders, two of which, the Finance Minister and the Minister of Justice, remained in office as if nothing happened and ran successfully for reelection a few months later. So did the former Prime Minister, who had resigned under public pressure after the scandal broke. Three and a half years later, nothing has been divulged about the movements in those Icelandic Panama accounts of the ministers or their tax treatment. The former Prime Minister now leads a growing party of the nationalist variety seen in many other European countries.

3 Crash Followed by Recovery

The rapid expansion of the banking system launched the economic boom in the early 2000s. In 2003, at the time of privatization, the banks had assets equivalent to less than twice the country's GDP. By the end of 2007 their assets had increased to eight times GDP. As described by Benediktsdóttir et al. (2011), the average annual asset growth from 2004 to 2008 was between 50 and 60% in the three main banks. The money was lent to limited liability companies, mostly in the form of foreign currency loans that were then invested in the domestic stock market, used to take over domestic firms or used to buy foreign firms, mostly in the U.K. and Denmark. About 80% of the borrowing never entered the country but was used to buy foreign assets. The holding companies earned high profits because of the stock market bubble that the credit expansion created and because of the appreciation of the Icelandic króna that lowered the domestic currency value of their debt. Hence, the foreign borrowing itself created profits by elevating the stock market as well as the currency. As described above, the good sovereign credit rating made the borrowing possible. At least with the benefit of hindsight, we can say that the combination of laws that allowed individuals to set up limited liability holding companies and borrow to finance investment in shares, the free flow of capital in the EU internal market that allowed foreign currency borrowing within Iceland and, last but not least, the willingness of banks to lend in foreign currencies to unhedged parties while discounting the credit risk created private incentives for credit expansion that then generated private profits while creating enormous financial risk for the banks and the sovereign. Perhaps the reason why the banks were willing to lend in foreign currencies at low interest rates had something to do with their owners being among the largest borrowers. Higher domestic interest rates did not curb the economic expansion because borrowing was mainly in foreign currencies. Rather, the high interest rates generated profit opportunities and attracted capital inflows that increased domestic demand, thus reducing the effectiveness of the inflation-targeting regime launched in 2001.

Monetary policy relied exclusively on high interest rates to curb the economic expansion and achieve an inflation target. However, the main effect of the high interest rates was to attract speculative capital, the carry trade, and to induce local borrowers to borrow in foreign currencies. Thus, domestic firms and foreign investors were long in the króna, the only difference being that the foreign investors could exit the carry trade more quickly. The resulting appreciation of the currency made the import sector and construction expand while exports suffered lower profits. The effect on relative prices, that is, the real exchange rate, and credit supply was on net expansionary. What was missing was a second instrument to stem the hot money inflows. This instrument could have taken the form of an increased general reserve requirement that would have curbed credit expansion in foreign currencies, an option vehemently opposed by the banks before the crash,[13] or a special reserve requirement on the investment of foreign investors in listed bonds, a measure adopted in 2016 when the carry trade started again. With two instruments, the central bank could have raised interest rates without triggering the capital inflows, the exchange rate channel of monetary policy would have been muted and the interest rate channel strengthened. However, the central bank did not attempt to reduce the capital inflows; on the contrary, the inflows were welcomed because they caused an appreciation, which made measured inflation fall, which supposedly helped the central bank attain its inflation target.[14]

Rising domestic interest rates did not reduce the rate of credit creation, which was mostly in foreign currencies. The credit-generated stock market bubble had both elements of Keynes (1936) and Minsky (1986).[15] It was not clear to most investors that a crash was bound to happen until it was too late to avert. Instead, investors and businesses developed a herd mentality, incapable of calculating any "true" probabilities of events to come. They each trusted each other's instincts. From what started out as profitable investments (mostly in other countries) that paid for the cost of financing, the borrowing gradually developed into a Ponzi scheme where banks and investors had to borrow from foreign banks to be able to roll over and pay interest on existing debt. The goings on were justified by storytelling of the kind exemplified by Exhibit C. However, as the realization dawned that the chicken would come home to roost, the owners of many an enterprise started to remove assets from the doomed,

[13]In fact, to accommodate the banks' wishes, the Central Bank reduced the reserve requirement before the crash.

[14]See Dooley (2019) and Ghosh et al. (2017), among others, on the use of selective capital flows.

[15]See Chap. 12 of Keynes's *General Theory*.

not yet bankrupt businesses. Even in the best of times, the banks' owners had borrowed from their own banks, often with bullet loans that only had to be repaid at the end of their maturity.

Exhibit C Delirium Praesidis
Recently, I have often found myself cornered at various functions, especially here in London, and pressured to explain how and why daring Icelandic entrepreneurs are succeeding where others hesitate or fail, to reveal the secret behind the success they have achieved.

It is indeed an interesting question how our small nation has in recent years been able to win so many victories on the competitive British, European and global markets, especially because for centuries we were literally the poorest nation in Europe, a community of farmers and fishermen who saw Hull and Grimsby as the main focus of their attention, a nation that only a few decades ago desperately needed to extend its fishing limit in order to survive, first to 12 miles, then to 50 and finally to 200 miles. Each time Britain sent the Navy to stop us but each time we won—the only nation on earth to defeat the British Navy, not once but three times. With this unique track record, it is no wonder that young entrepreneurial Vikings have arrived in London full of confidence and ready to take on the world!

The track record that Icelandic business leaders have established is also an interesting standpoint from which to examine the validity of traditional business teaching, of the theories and practice fostered and followed by big corporations and business schools on both sides of the Atlantic. It enables us to discuss the emphasis on entrepreneurial versus structural training, on process versus results, on trust versus career competition, on creativity versus financial strength.

I have mentioned this morning only some of the lessons which the Icelandic voyage offers, but I hope that my analysis has helped to clarify what has been a big mystery to many. Let me leave you with a promise that I gave at the recent opening of the Avion Group headquarters in Crawley. I formulated it with a little help from Hollywood movies: "You ain't seen nothing yet".

Excerpts from HOW TO SUCCEED IN MODERN BUSINESS: LESSONS FROM THE ICELANDIC VOYAGE, A speech by the President of Iceland Ólafur Ragnar Grímsson at the Walbrook Club London 3rd May 2005. https://agbjarn.blog.is/users/fa/agbjarn/files/05_05_03_walbrook_club.pdf

Black (2005) describes how banks can be looted from within when they grow fast, make bad loans at high interest rates, are highly leveraged and have low bad-debt reserves. Akerlof and Romer (1993) describe how implicit state backing can help bankers to profit from bankruptcies by borrowing and taking in deposits, paying themselves dividends that are greater than the net worth of the business and then leaving it to the taxpayer or to creditors to pay off the debt. This did happen in Icelandic banks. Apart from paying out dividends, there are several other ways of exploiting deposit insurance and implicit state backing such as by making high-risk loans, buying highly leveraged firms and creating value through an accounting goodwill and paying high salaries both in the banks and in businesses so acquired from an overstated accounting income.

Akerlof and Romer's (1993) description of the financial crisis that hit Chile in 1982 resembles what happened in Iceland in 2008. The banks and their owners exploited the interest differential between domestic and foreign (dollars) currency bonds to create accounting profits that were used to justify the paying of dividends to the banks' owners. The banks did not increase the allowance for bad debt despite having generated foreign-currency denominated loans to unhedged households and businesses. Both in Chile and in Iceland the owners of the banks often happened to be the owners of the companies that received the biggest loans, which ensured that all the gains from such transactions went to these owners. In effect, the profits were generated by the underreporting of risk, something that was legal at the time and benefitted the owners but endangered financial stability.

While the bankers had created serious risks to domestic financial stability by borrowing from foreign banks backed by the sovereign's credit ratings, the government itself decided to deflect the risk to the foreign creditors by allowing the debt of the banking system to outgrow the ability of the state to support the solvency of the banks in a crisis and the ability of the central bank to provide liquidity support. In the end, foreign creditors would carry the losses incurred in case of collapse while the bankers would reap the benefits from the survival of the banks. The net investment position of Iceland belied its gross indebtedness. At the time of the collapse, the foreign assets exceeded 500% of GDP while the foreign debt was close to 600% of GDP rendering the net position negative by roughly 100% of GDP. It was claimed that the assets were undervalued so that there actually was no problem. However, when the crisis hit, it was gross debt, not net debt, that was important and its size made it impossible for the central bank or the sovereign to come to the rescue.

Pursuing the earlier comparison with Denmark a little further, Iceland caught up as the twentieth century wore on, briefly reaching economic parity with Denmark in per capita terms just before Iceland's financial crash of 2008 temporarily separated the two again (Fig. 5). The effect of the bubble economy from 2003 to 2008 is visible in Fig. 5 but note that after the bursting of the bubble the economy continued to grow along its previous path. A more revealing comparison, however, is per capita income per hour worked, a measure that takes into consideration the work effort behind the national output. Icelanders worked 1500 h per worker during 2017 compared with 1400 h in Denmark and 1700 h in Italy (Conference Board 2019). In Iceland, long-standing economic inefficiency caused by, *inter alia*, insufficient competition in agriculture, banking, fisheries, and so on, has long kept real wages lower than in Denmark, inducing employees to work longer hours and stay longer in the labor force than they might otherwise do (Gylfason 2015). This helps to explain why

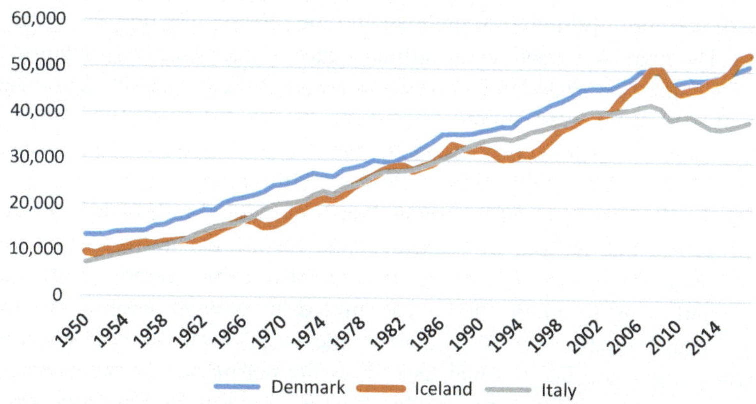

Fig. 5 GDP per capita 1950–2017 (2018 USD) (Source: The Conference Board Total Economy Database)

output per hour worked in 2017 was USD 63 in Iceland compared with USD 72 in Denmark and USD 54 in Italy (same source).

4 Incentives and Morals Gone Wrong

We have described how individuals and firms had private incentives to profit that created the conditions for a financial crisis. What were these conditions and institutions that made it profitable to do things that were ultimately harmful for the economy? By answering this question, we can better understand the more profound question posed at the beginning of this study: Why do destabilizing capital flows affect some countries while others manage to escape?

We identify five main factors.

The first part of the answer comes from the decision by the authorities to make Iceland an international financial center. These changes amounted to relaxing requirements already in place in Iceland without violating the minimum requirements of the EU Directives. These changes included increased authorization for banks to invest in non-financial businesses, to extend credit to directors, to invest in real estate and real estate companies, to lend money to buy own shares and to own other credit institutions in addition to less stringent requirements concerning the permission of securities companies to operate insurance companies (Benediktsdóttir et al. 2011).

In effect, the changes allowed the banks to increase their capital through accounting gimmicks, something that was needed for their rapid growth. The banks bought their own shares in order to manipulate the share price. Since they could only own 10% of the value of outstanding shares, the banks gradually sold the shares they had previously bought to selected customers and employees. These purchases were financed by the banks by granting the buyers of the shares bullet loans where a payment of the entire principal of the loan and the interest was due at the end of the loan term. The loans were defined as ordinary bank loans instead of unclaimed share capital. The capital generated in this way was around 40% of market capitalization at the end of 2007 and even more at the time of the collapse. When a financial institution increases equity, it is essential that the funding come from outside the financial institution lest equity become illusory (Hilmarsson and Svavarsson 2018). Outside Iceland, shares cannot be counted as bank capital unless they are paid for with money. For this reason, it is unacceptable for a financial institution to count unclaimed share capital as core capital because it does not create a buffer against losses. To add insult to injury, the banks counted the interest on the bullet loans as revenues, hence overstating profits since the interest payments were never collected.

A second important institutional factor was the permission banks were given to lend in foreign currencies to unhedged businesses operating in domestic currency as well as to households with incomes only in domestic currency. This was in accordance with EEA directives. This allowed the banks to borrow from foreign banks and lend to domestic parties in foreign currencies thus maintaining a foreign exchange balance on their books while creating a currency mismatch for their customers. This in effect made businesses take a long position in the domestic currency and a short position in foreign currencies. The bet paid off while the currency appreciated during the period of the capital inflow until spring 2008. But when a sudden stop of the capital flows made the currency tank almost all businesses, financial as well as non-financial, became insolvent and hence also the banks. Of course, prudent banks in other EEA countries would have added a risk premium to the foreign rate of interest when lending to unhedged parties, a premium that reflected the risk of default. But prudence is put aside when dealing with friends or oneself.

Because of the prevalence of loans denominated in foreign currencies businesses that would have preferred to avoid the foreign currency risk were forced to borrow in foreign currency because they would otherwise have had significantly lower profitability. The privately desirable financing was in foreign currencies although this was clearly not socially desirable, as experience would show.

A third factor conducive to the risk-seeking behavior was the supervisor's failure to prevent banks from lending to related parties. The owners created a complicated ownership structure so that it was often difficult to trace the ownership of banks to the ownership of different non-financial companies. By not seeing through this cobweb, the supervisor allowed the banks to lend to the owners, which in effect amounted to the owners borrowing from foreign banks at lower rates due to the implicit government backup and making profits from speculation and misleading accounting. To quote the Special Investigation Commission's (2010) discussion of the father and son that were the main owners of Landsbanki: "When Landsbanki collapsed, Björgólfur Thor Björgólfsson and companies affiliated to him were the bank's largest debtors. Björgólfur Guðmundsson [his father] was the bank's third largest debtor. In total, their obligations to the bank … [were] higher than Landsbanki Group's equity."

Fourth, the authorities allowed the banks to become too big to save, hence making them inherently unstable as shown by Diamond and Dybvig (1983). In addition, when collecting foreign deposits in the U.K. and the Netherlands, one of the three large commercial banks (Landsbanki) could collect deposits that exceeded the foreign currency reserves of the central bank by a factor of five. Last but not least, the rapid growth of the banks was powered in part by mania of the kind described by Kindleberger and Aliber (2005) as well as by Keynes (1936) and Minsky (1986). Ambition exceeded ability and hope exceeded reality. Extremely rapid credit growth for years on end is always and everywhere a harbinger of declining creditworthiness. Paid by the banks, the rating agencies failed. The voices of those few outsiders[16] who warned against excessive expansion and advocated for higher reserve ratios to rein in the banks were drowned by those who, like the finance minister, asked dismissively: "Don't you see the party, guys?" Without exception, insiders kept their silence (Exhibit D).

As in other bubble economies in the past, a story emerged that justified the booming economy. The story in Iceland was of financial liberalization unleashing entrepreneurship and business acumen inherent in local culture. Before the crash, the president of the country claimed that the Icelandic success story would change the way business schools on both sides of the Atlantic would design their curricula (recall Exhibit C). These stories served to justify the good times, to help underpin confidence in continued prosperity and to sideline those who had doubts. The owners of the banks also owned most of the media, which extolled the wonders of the bank expansion.

Exhibit D From Procrastination to Obedience

Akerlof (1991) shows how procrastination leads to undue obedience due to myopic and time-inconsistent optimization, using his analysis to derive suboptimal outcomes from individually rational behavior. This, says Akerlof, is why smokers find it more convenient to stop smoking tomorrow rather than today. And this is why the Russian *nomenklatura* found it opportune not to stand up to Stalin until after his death, and also why opponents of the Vietnam War within the Johnson Administration, Bill Moyers and others, did not protest against the war in public but chose instead one after another to leave the administration relatively quietly.

What does this have to do with Iceland's financial collapse? In sworn testimony before the Court of Impeachment in 2012, senior officials at the Central Bank of Iceland stated that it had become clear in 2006 that the

(continued)

[16]The outsiders include Icelandic and foreign university professors, some of whose pre-crash writings appear in Aliber and Zoega (2011). Many other academics and journalists stayed silent, however, or cluelessly towed the line of the banks, the business community, and the government. Some still do. For more, see Gylfason (2015, 2019).

commercial banks were operated like a Ponzi scheme and could not be saved. Even so, the Central Bank continued to lend them money for two more years, culminating in a EUR 500 million loan to Kaupthing on 8 October 2008 that virtually emptied the Central Bank's foreign exchange reserve. In a telephone conversation that was not leaked until 2017, the governor told the prime minister: "We will not get the money back." [Our translation.] The prime minister who now represents the Nordic and Baltic constituency on the World Bank's Executive Board recently expressed regret that he did not know where the money went. In April 2011, however, it was reported by *Viðskiptablaðið*, an Icelandic newspaper, that a third of the EUR 500 million was deposited the same day in an account in Tortola, a Caribbean tax haven.[17] The statute of limitations ran out in October 2018.

No Central Bank official spoke up or resigned in protest to warn the public of the goings on. Bill Moyers knew that he could find employment elsewhere. Perhaps the Icelandic Central Bank officials who saw through the Ponzi game could not be so sure. Further, as John Kenneth Galbraith wrote in his book *The Great Crash 1929* (1988, p. 160): ". . . the action will always look, as it did to the frightened men in the Federal Reserve Board in February 1929, like a decision in favor of immediate as against ultimate death. As we have seen, the immediate death not only has the disadvantage of being immediate but of identifying the executioner."

Overall, the institutional setup we have described and the mindset that propelled it created incentives for private gain at the expense of financial stability. The owners of the banks could borrow from foreign banks at lower interest rates because the banks were systemically important, lend to themselves in foreign currencies at low interest and invest in the domestic high-interest-rate currency, buy foreign businesses or take over a variety of firms that enjoyed market power in the local market. The bankers profited while imposing the risk on the taxpayer, the sovereign's credit rating and foreign creditors.

In addition, business ethics did not involve concern about responsible behavior. They were formed by a blind belief in the invisible hand of the market without regard to the need for prudent oversight and they owed more to Adam Smith's *Wealth of Nations* than to his *Moral Sentiments*, which explains why people may often behave in a way that takes the interest of others into account as in modern behavioral economics. In effect, the business community had convinced itself that a market economy would make their reckless behavior serve the national interest without realizing all the assumptions about the number of participants in the market, perfect information and market completeness necessary for this outcome (Bowles 2016).

[17] See https://www.vb.is/frettir/lindsor-var-stjornad-af-stjornendum-kaupthings/62474/

5 Conclusion

We see Iceland's financial collapse of 2008 as a clear case of the failure of collective action, that is, poor governance. Each agent could contribute to financial stability, knowing that no one gains from a financial crisis except those who get away with the loot. However, the behavior of one agent is in most cases not important for financial stability so one person or one firm could always start borrowing in foreign currencies, borrowing to buy shares or lend to related parties. Problems arise when almost all do this. The solution proposed by Olson (1971, 2000) was for a government to use its persuasive or coercive power to make individuals and firms behave in a more responsible manner, to follow the rules. However, for this to be possible the rules must be seen to be just and fair and be seen to ensure the stability of the system in our context. But when the government sets rules that facilitate private gain at public expense, it invites looting and kleptocracy. There is no evidence that the financial authorities in Iceland tried to ensure stability rather than to facilitate the financiers′ profiteering. Using Olson's terminology, the government represented special interests, which prevented it from pursuing the encompassing interests that took the welfare of the whole of society into account. A government with fully encompassing interests would have aligned private interests with social interests and stepped in to stop the bankers before it was too late. A government too close to the bankers did not.

The core of the problem, in addition to the unstable international monetary system and the dangers of free capital flows, was a failure of the state. This failure can be traced partly to incompetence, including a lack of basic understanding of banking in the shape it took, but partly also to a culture of rent seeking, economic concentration and corruption, including criminal behavior, as we have described. It seems that the pervasiveness of rent seeking, in particular, has been such that the authorities almost take it for granted that economic rent be allocated to cronies even if their reckless behavior remains a threat to financial stability. A powerful state would have passed regulations and laws that induced private agents to do what was best for the country—maximizing the total value added—but the state was weak. The necessary rules were not put in place and laws passed that would have prevented some financiers from benefitting at the expense of others. The weakness of the state and its organs and the massive wealth of the financiers led society astray as agents who were often rational about their own self-interest although sometimes blinded by optimism and euphoria made decisions that led to the abyss.

References

Ades, A., & Di Tella, R. (1997). The new economics of corruption: A survey and some new results. *Political Studies, 45*, 496–515.

Ades, A., & di Tella, R. (1999). Rents, competition and corruption. *American Economic Review, 89*, 982–993.

Aidt, T. S. (2003). Economic analysis of corruption: A survey. *Economic Journal, 113*(491), F632–F652.
Akerlof, G. A. (1991). Procrastination and obedience. *American Economic Review, 81*(2), 1–19.
Akerlof, G. A., & Romer, P. M. (1993). Looting: The economic underworld of bankruptcy for profit. *Brookings Papers on Economic Activity, 2*, 1–73.
Alesina, A., & Spolaore, E. (2003). *The size of nations*. Cambridge, MA: MIT Press.
Aliber, R. Z. (2016). A Lego approach to international monetary reform. *Atlantic Economic Journal, 44*(2), 139–157.
Aliber, R. Z., & Zoega, G. (Eds.). (2011). *Preludes to the Icelandic financial crisis*. New York: Palgrave Macmillan.
Benediktsdóttir, S., Daníelsson, J., & Zoega, G. (2011). Lessons from a collapse of a financial system. *Economic Policy, 26*, 183–235.
Benediktsdóttir, S., Eggertsson, G. B., & Thórarinsson, E. (2017). The rise, fall, and resurrection of Iceland: A postmortem analysis of the 2008 financial crisis. *Brookings Papers on Economic Activity, 2*, 191–281.
Black, W. K. (2005). *The best way to rob a bank is to own one: How corporate executives and politicians looted the S&L industry*. Austin: University of Texas Press.
Bowles, S. (2016). *The moral economy: Why good incentives are no substitute for good citizens*. New Haven: Yale University Press.
Conference Board Total Economy Database™. (2019, April). Retrieved from https://www.conference-board.org/data/economydatabase/index.cfm?id=25667
Diamond, D. W., & Dybvig, P. H. (1983). Bank runs, deposit insurance, and liquidity. *Journal of Political Economy, 91*(3), 401–419.
Dooley, M. (2019). Prudential regulation and capital controls. In R. Aliber & G. Zoega (Eds.), *The 2008 global financial crisis in retrospect* (pp. 43–48). Cham: Palgrave Macmillan.
Galbraith, J. K. (1988). *The great crash 1929*. Boston: Houghton Mifflin Company. Original edition 1954.
Gallup. (2013). "Government corruption viewed as pervasive worldwide" by Jan Sonnenschein and Julie Ray. Retrieved October 1, 2018, from http://www.gallup.com/poll/165476/government-corruption-viewed-pervasive-worldwide.aspx
Gallup. (2018). *Traust til stofnana* [Trust in institutions]. Retrieved October 2, 2018, from https://www.gallup.is/nidurstodur/thjodarpuls/traust-til-stofnana/
Ghalia, T., & Fidrmuc, J. (2015). The curse of tourism? *Journal of Hospitality & Tourism Research, 42*(6), 979–996.
Ghosh, A. R., Ostry, J. D., & Qureshi, M. S. (2017). *Taming the tide of capital flows: A policy guide*. Cambridge, MA: MIT Press.
Gylfason, T. (1999). Exports, inflation, and growth. *World Development, 27*(6), 1031–1057.
Gylfason, T. (2009, August 19). Is Iceland too small? *VoxEU*.
Gylfason, T. (2013). From collapse to constitution: The case of Iceland. In L. Paganetto (Ed.), *Public debt, global governance and economic dynamism* (pp. 379–420). Milan: Springer-Verlag.
Gylfason, T. (2015). Iceland: How could this happen? In T. M. Andersen, M. Bergman, E. Svend, & H. Jensen (Eds.), *Reform capacity and macroeconomic performance in the Nordic countries* (pp. 310–348). Oxford: Oxford University Press.
Gylfason, T. (2019). Ten years after: Iceland's unfinished business. *In Aliber and Zoega, 2019*, 297–235.
Gylfason, T., & Wijkman, P. M. (2016). Double diversification with an application to Iceland. In S. Mahroum & Y. Al-Saleh (Eds.), *Economic diversification policies in natural resource rich economies* (pp. 267–294). London: Routledge.
Gylfason, T., & Zoega, G. (2018). The Dutch disease in reverse: Iceland's natural experiment. In L. Paganetto (Ed.), *Getting globalization right: Sustainability and inclusive growth in the post Brexit age* (pp. 13–36). Berlin: Springer.

Hilmarsson, J. T., & Svavarsson, S. (2018). *Equity out of nothing*. Retrieved March 13, 2019, from https://notendur.hi.is/gylfason/Final%20Equity%20out%20of%20nothing%20II.pdf

Ingimundarson, V. (1996). *Í eldlínu kalda stríðsins. Samskipti Íslands og Bandaríkjanna 1945–1960* (In the Crossfire: Iceland, the United States, and the Cold War, 1945–1960). Reykjavík: Vaka-Helgafell.

International Monetary Fund. (2009, October). IMF country report no. 09/306 Iceland: Staff report for first review under stand-by arrangement and requests for extension of the arrangement, waivers of nonobservance of performance criteria, and rephasing of access. Retrieved July 30, 2019, from https://www.imf.org/external/pubs/ft/scr/2009/cr09306.pdf

Jain, A. K. (2001). Corruption: A review. *Journal of Economic Surveys, 15*(1), 71–121.

Jensdóttir, J. S. (2017). Ákærur og dómar vegna hrunmála (Indictments and verdicts in crash-related cases), Gagnsæi (Transparency), Samtök gegn spillingu (Alliance against corruption). Retrieved October 8, 2018, from http://www.gagnsaei.is/2017/12/29/domar1/

Jóhannesson, S. (2015, April 11). *Verðmæti auðlinda í almannaeigu* [The value of publicly owned natural resources]. Conference presentation at the University of Iceland.

Júlíusson, T. S. (2018). *Kaupthinking*. Reykjavík: Bjartur.

Keynes, J. M. (1936). *The general theory of employment, interest and money*. London: Macmillan.

Kindleberger, C., & Aliber, R. Z. (2005). *Manias, panics and crashes: A history of financial crises*. London: Palgrave Macmillan.

Krueger, A. (1974). The political economy of the rent-seeking society. *American Economic Review, 64*(3), 291–303.

Laeven, L., & Valencia, F. (2013). Systemic banking crises database. *IMF Economic Review, 61*(2), 225–270.

Maddison, A. (2019). *Maddison historical statistics*. Retrieved from https://www.rug.nl/ggdc/historicaldevelopment/maddison/

Minsky, H. P. (1986). *Stabilizing an unstable economy*. New York: McGraw Hill.

Olson, M. (1971). *The logic of collective action: Public goods and the theory of groups, second printing*. Cambridge, MA: Harvard University Press.

Olson, M. (2000). *Power and prosperity: Outgrowing communist and capitalist dictatorships*. New York: Basic Books.

Panama Papers. (2016). *The panama papers: Exposing the rogue offshore finance industry*. Retrieved July 9, 2019, from https://panamapapers.icij.org/

Pendergast, S. M., Clarke, J. A., & Van Kooten, G. C. (2011). Corruption, development and the curse of natural resources. *Canadian Journal of Political Science/Revue canadienne de science politique, 44*(2), 411–437.

Reinhart, C. M., & Rogoff, K. S. (2009). *This time is different: Eight centuries of financial folly*. Princeton, NJ: Princeton University Press.

Reinhart, C. M., & Rogoff, K. S. (2014). Recovery from financial crises: Evidence from 100 episodes. *American Economic Review: Papers & Proceedings, 104*(2), 50–55.

Van Rijckeghem, C., & Weder, B. S. (2001). Corruption and the rate of temptation: Do low wages in the civil service cause corruption? *Journal of Development Economics, 65*, 291–307.

Rose-Ackerman, S. (Ed.). (2006). *International handbook on the economics of corruption*. Northampton, MA: Edward Elgar Publishing.

Rose-Ackerman, S., & Søreide, T. (Eds.). (2011). *International handbook on the economics of corruption, volume two*. Northampton, MA: Edward Elgar Publishing.

Shleifer, A., & Vishny, R. W. (1993). Corruption. *Quarterly Journal of Economics, 108*(3), 599–617.

Special Investigation Commission. (2010). Report of the Special Investigation Commission (SIC). Report delivered to the Icelandic Parliament 12 April. Retrieved October 4, 2018, from https://www.rna.is/eldri-nefndir/addragandi-og-orsakir-falls-islensku-bankanna-2008/skyrsla-nefndarinnar/english/

Tanzi, V. (1998). Corruption around the world, causes, consequences, scope, and cures. *IMF Staff Papers, 45*(4), 559–594.

Thorláksson, I. H. (2015). *Veiðigjöld 2015. Annar hluti* [Fishing fees 2015. Part two].
Transparency International. (2018). Corruption perceptions index 2017. Retrieved October 8, 2018, from https://www.transparency.org/news/feature/corruption_perceptions_index_2017
Treisman, D. (2000). The causes of corruption: A cross-national study. *Journal of Public Economics, 76*, 399–457.

Towards a New Taxonomy of Manufacturing Countries

Livio Romano and Fabrizio Traù

Abstract The scope of this paper is to propose a novel approach to the categorization of manufacturing development, aimed at accounting for the major global transformations that have occurred in the organisation of industrial activity in the last decades. It first addresses the way manufacturing development can be defined in order to provide a measure of the degree of industrialization of different countries, and then suggests a new taxonomy accordingly. Attention is paid to the fact that in the course of time countries can—and usually do—move from one group of manufacturers to another. Moreover, it is shown that cross-country differences in the degree of industrialisation are also mirrored by differences in their institutional features. Results offer some important lessons for industrial policy.

Keywords Industrial development · Taxonomy · Institutions

JEL Classification F63 · O14 · O25

1 Introduction

This paper deals with the problems of measuring manufacturing development as well as with interpreting its actual meaning. Both issues have been heightened by the major changes that have affected the organisation of industrial activity all across the world in the years of globalisation, which have made obsolete the view according to which the world was sharply divided into two parts, largely unrelated to each other: on the one side, the 'North' of the world, basically coinciding with West Europe,

L. Romano (✉) · F. Traù
Centro Studi Confindustria, Rome, Italy
e-mail: l.romano@confindustria.it; f.trau@confindustria.it

L. Paganetto (ed.), *Capitalism, Global Change and Sustainable Development*, Springer Proceedings in Business and Economics,
https://doi.org/10.1007/978-3-030-46143-0_12

North America and Japan; on the other, the 'South', including almost all other economies.[1]

Indeed, the last two decades have led to an unprecedented degree of productive interdependence at the country level, linking together economic systems originally characterised by strong differences in their degree of industrialisation and allowing the emergence of new manufacturing powerhouses. Not only has the bursting onto the economic scene of these players brought about the downsizing of manufacturing production within the 'North', but it has also been accompanied by the persistent stagnation of a conspicuous part of the economies of the 'South' of the world—determining in turn an increase in inequality within the whole group of laggards. This has led to an overall reshaping of the landscape of global manufacturing, implying a change in the relative distance among all the countries involved in manufacturing activity.

In the face of such deep transformations we still lack a metrics which is able to return a consistent picture of what global manufacturing has actually become. In fact, the standard way the degree of industrialisation of different countries is looked at is still largely rooted in the 'North-South' dichotomous way of thinking about development, and appears increasingly at odds with reality.

The contribution of this paper is to propose a novel approach to the categorization of countries, which rests on economic theory and empirical research to build a multi-dimensional metrics for measuring the degree of industrialisation of a given country. In particular, by applying cluster analysis to a set of complementary structural economic indicators it is possible to account for the increasing cross-country heterogeneity in the characterisation of a manufacturing base and to shed new light on the different directions undertaken in the last decades by manufacturing development worldwide.

The paper is organised as follows: Sect. 2 provides an overview and an interpretive framework of the above-mentioned changes in world manufacturing, discussing the conditions for industrial development to happen and the reasons for heterogeneous paths to emerge. Section 3 addresses the way manufacturing development can be defined in order to provide a measure of the degree of industrialiszation of different countries. Section 4 first describes the methodology employed to define a new taxonomy of manufacturing countries, and then shows the results of its implementation, illustrating the specific features which characterise each of the resulting tiers (or groups) of manufacturers. Attention is paid in this connection to the fact that in the course of time countries can—and usually do—move from one tier of manufacturers to another. Section 5 focuses on the role played by institutional factors in explaining different degrees of development, revealing important differences across tiers of manufacturers. Section 6 discusses the policy implications and concludes.

---

[1]This framework has set apart for many years, as a separate group, the ensemble of the East European countries, which were indeed characterised by a quite strong role of the industrial sector in the economy but were nevertheless classified outside the boundaries of the industrialised world.

2 Industrial Development in the Globalisation Age

2.1 Organising Production Activity in the Global World

The Globalisation Age has coincided with the expansion of manufacturing beyond the G7's borders and the emergence of new countries as manufacturers.[2] Indeed, the view according to which the world had to be intended in dichotomous terms as to industrialisation had long been challenged by the emergence of the four 'Asian tigers' since the 70s, that set something new somewhat in between the 'North' and the 'South'. Yet this did not translate into a rethinking of the idea according to which the world was to be divided into two blocs, until the emergence in the new century of *large* manufacturing countries such as China (and, to a lesser extent, the other BRICs) did actually alter the whole mechanism of industrial development at the global level.

This happened due to the very size of the new emerging countries (whose impact became suddenly visible), and the fact that their development hinged upon an increasing division of labour on a cross-country basis, creating a widespread flow of exchanges on productive grounds between the two sides of the world. Where did this came from, and how did it work?

A major event changing the basic conditions for industrial development in a global perspective has been the 'unfreezing' of large economic systems that still in the early 90s were sealed within a parallel world, outside of the area of market exchanges. Within an exceptionally narrow period of time (the decade spanning from Deng's launch of the four modernisations in 1978 to the fall of the Berlin wall in 1989), new economic areas (and populations) *simultaneously* came to be part of the global world, sharply widening the boundaries of the world prone to market rules.

The outcome was a one-off event, unprecedented in industrial history, that almost suddenly opened up to the 'North' a new labour market characterised by an unlimited availability of labour at extremely low cost, linking together the most industrialised and the less industrialised economies in the world.[3] This new order, outlining a situation close—on a global scale—to that described in the Lewis-Kindleberger's model (Lewis 1954), has set the premises for the launch in underdeveloped countries—*wherever some manufacturing know-how was already available* (infra)—of an industrialisation process, avoiding for a very long time any tension on output prices.

Yet industrial development did not happen anywhere: what is the mechanism whereby such a process did actually take place and why did it show different intensities in different areas? The basic point here is that as far as in laggard countries domestic demand was bounded to a minimum, the way out from underdevelopment

[2]For a wider treatment of the subject see Traù (2016).

[3]The phenomenon, including India and other Asian underdeveloped economies, has been termed by Freeman (2007) as 'great doubling'.

had to come from external demand. In this connection the head start comes from what was called the second 'great unbundling', enabling the exploitation on a global scale of the so-called trade in tasks.[4] Put in a few words, the phenomenon has consisted in the transfer abroad of a part of domestic production of industrial countries. As far as it became possible to exploit huge cross-country cost differentials in underdeveloped countries, the conditions were created for a more or less extensive transfer of production stages—in general the more labour intensive ones—from the former to the latter.

Through this type of cross-country diffusion of production activities, supply chains became fragmented on a global scale.[5] On the whole the fragmentation of supply chains at the global level has determined an unprecedented demand shock. The breakdown of (some) complex industrial processes into separated production phases was the channel through which many economies had the possibility of being involved in the production of specific components.[6] This was the channel whereby emerging countries could gradually develop—starting by initially competing on prices—economies of specialisation that could lead in time to the unfolding of growing dynamic returns. Where did this happen, and where did it not?

2.2 *The Economic Conditions for Emerging*

Following Alice Amsden (2001), we can say that at the dawn of globalisation the laggards were already divided into two groups: on the one hand the 'Rest' (namely the group of countries oriented, to a greater or lesser extent and with differing degrees in the process implementation, towards an explicit industrialisation objective), and, on the other, the 'Remainder' (lacking manufacturing knowledge and adequate institutions).[7] In Amsden's own words, "[c]ountries in 'the rest' that industrialised rapidly after World War II had accumulated manufacturing experience in the pre-war period. This differentiated them from countries in 'the remainder'".

[4]See, in particular, Baldwin (2006, 2012, 2014) on the former and UNIDO (2009), WTO and IDE-JETRO (2011) on the latter.

[5]The endless literature on Global Value Chains (GVCs) cannot be summarised here. Basic reference include, among others, Gereffi et al. (2005), Sturgeon (2008), Nolan et al. (2008), Cattaneo et al. (2010), and Gereffi (2014).

[6]This was very much enhanced by the lifting of barriers to commercial integration following the new globalisation paradigm.

[7]According to Amsden (2001), the 'Rest' includes China, India, Indonesia, Korea, Malaysia, Taiwan and Thailand in Asia; Argentina, Brazil, Chile and Mexico in Latin America, and Turkey in the Middle East. The development of these countries, that Amsden (see in particular Chaps. 4 and 5) illustrates through the review of a massive series of analyses relating to them individually, takes on a different profile depending on the way in which the economies lagging behind acquired manufacturing knowledge (and, in particular, depending on the presence or absence of a prior colonial experience).

The point here is that "[p]ath dependence was such that no economy emerged from the blue as an industrial competitor" (2001, p. 99).

As a result of the initial differences among the different areas of the developing world, then, industrialisation took place only in some developing countries and not in all of them. This set the very premises of what had to happen later, in the course of the Globalisation Age: for the emergence of new global manufacturers to unfold, the prior existence of an adequate endowment of manufacturing knowledge, as well as a peculiar institutional setting, was required.

Of course, the boundaries of the group of countries that in the following have been actually involved in an industrialisation process are somewhat larger than those strictly corresponding to the 'Rest'. This is due to the very logic of industrial development itself, which stems from a gradual strengthening of (cumulative) backward and forward linkages in the sense suggested by Hirschman (1958, 1981)—i.e. from the activation of an endogenous widening of the supply matrix following the expansion of demand. In this perspective the 'Rest' itself has played an active role in the development of neighbouring economies, adding in turn new external demand to that coming from the 'North' and allowing other laggards to enter new manufacturing activities. This has proved to be most evident in Eastern Asia, where a new integrated area has gradually emerged, including, other than Japan and the 'Rest' economies (China, Indonesia, South Korea, Malaysia, Taiwan and Thailand), laggards such as Philippines and Vietnam.

A further widening of the boundaries of emerging economies has been fed by the transition of former East European planned economies to the market. In this respect the previous existence of a widespread manufacturing knowledge in such countries has paved the way to the development and gradual thickening of market linkages between them and the countries belonging to the European Union (maxime Germany).[8] With the notable exception of Russia, this has brought about a sudden passage of some of these countries—previously segregated elsewhere—into the international manufacturing system, making them an important component of the global value chains mechanism.[9]

In which terms, and to what extent, have such transformations changed the industrial world?

[8]As to Germany, this has been claimed to involve a strong change in the organisation of manufacturing activity even within the country itself, in terms of the emerging of what has been called the German 'Bazaar economy' (Sinn 2006).

[9]Russia represents a spectacular example of the destructive consequences of an ideological application of the Washington Consensus rules, involving the simultaneous implementation of policies on price liberalization, accelerated privatization, tax provisions and commercial opening—without building up market institutions beforehand. The most evident counter-example is the maintaining of public property and control over domestic financial flows and capital movements in China, and in general the set of strategies pursued by the East-Asian economies (Lin 2009).

2.3 Heterogeneous Paths

The emerging industrial world has widened. Yet, this has not been a linear process. On the one side, the whole range of South American economies has experienced in the years of globalisation just an opposite situation: once abandoned the deprecated import substitution industrial (ISI) policies, both the rhythm of manufacturing development and the widening of the supply matrix have dramatically fallen instead of rising, coinciding with a relative marginalisation of the industrial system in such countries notwithstanding the parallel fragmentation of value chains on a global scale.[10] The lack of competitive advantages, following the absence, in the ISI years, of a policy aimed at developing dynamic increasing returns, has pulled these countries away from taking advantage of the widening of demand at the international level brought about by market liberalisation in the face of globalisation, unlike it happened in many East Asian countries.

On the other side, entering industrialisation via trade in tasks does not automatically mean entering the road to self-sustaining (endogenous) industrial development. The point here is that the globalisation logic itself tends to hinder the spread of manufacturing to a large range of activities, insofar as it makes crucial the possession of static comparative advantages. It turns out that in a context of strong international competition laggard economies have found, with few exceptions, heavy difficulties in achieving a significant degree of expansion of their supply matrix and hence of their industrialisation rate, facing an early upper bound to manufacturing development (Romano and Traù 2017).

This framework is furtherly complicated by the huge differences which characterise emerging countries as to their size. In general, whereas smaller economies necessarily need to base their development on specialisation (few industries that are competitive at the international level), larger ones can benefit from a domestic market that can activate supply also in non-competitive environments, therefore widening the absolute size of their manufacturing sector. This is what has allowed countries like Russia or Brazil to be considered as industrial countries, ranking within the world top 15 manufacturing producers, even in the face of a quite low industrialisation rate (as against population).

This means that the very scale of emerging manufacturers has been crucial in determining the role they have been able to assume within the international industrial system (actually, this is why the growth of small East Asian NICs along the 70s did not substantially alter the way the spreading of industrial development across the world was perceived). In particular, an important consequence of such a feature is the displacement that the more competitive (and large) emerging economies could exercise upon other weaker (and small) emerging ones—other than upon the

[10]From this point of view a different way has been followed by Mexico, as far as it succeeded in becoming part of the North-American supply chain.

industrialised world, which is what has typically caught the attention of applied economists over the last 20 years.[11]

In this overall context it also happens that the very functioning of the GVCs mechanism has gradually changed, asking for increasingly evolved suppliers and a greater concentration in upstream markets. This sets the premises for demand from abroad to become more selective. Hence, it may happen that a divide between the emerging economies that succeed in keeping the pace and those that do not is going to take place, thwarting the possibility for some countries to maintain their connection to international value chains.

Looking forward, the combination of these elements suggests that a new cleavage, in addition to the one that emerged in the past between the 'Rest' and the 'Remainder', may be dawning *within* the emerging world: some economies may be subject to falling behind and losing ground, whereas others may have the possibility to come closer to already industrialised countries.

Yet, the Globalisation Age has brought about relevant changes also within the industrialised world. The pivot of these changes has been the abovementioned mechanism whereby the two sides of the world have become mutually interdependent, entering a common future through the setting up of systematic market relations. For unbundling to occur at the international level, industrial structure in the "North" was to be already fragmented along vertical lines. From this point of view, an important role was played by the fact that many Western industrial countries had already experienced a process of vertical dis-integration in the course of the last quarter of the twentieth century, led by the search for flexibility in the face of higher uncertainty and rising competition (Traù 2003). Such a process started developing within national boundaries, in a few years largely trespassed them (Feenstra 1998).[12]

This mechanism has brought about a substantial cross-country reallocation of the supply potential. It is worth stressing that such a reallocation has to be intended in principle as a relative phenomenon, in the sense that, more than the result of a simple transfer from the industrialised countries to the emerging world, it is the outcome of different rates of growth in the different areas. In this sense, off-shoring has first to be viewed as the development abroad of new facilities and plants by multinational enterprises. Yet, in a great many cases the phenomenon has also involved a real cross-country transfer of productive activities (de-location). In such cases it has wrought a whittling down in the overall size of the manufacturing sector, as well as an uprise in import flows from emerging countries.

[11]The displacement can take the form of competition in trade—both direct (export vs. import) and indirect (on third markets)—as well as that of a diversion of the FDI flows coming from the 'North' from weaker emerging areas to stronger ones.

[12]It is worth stressing that this sorts of vertical fragmentation relates to a different issue with respect to the horizontal fragmentation (sub specie of intra-trade) that had long characterised the trade relations among industrialised countries, mainly—albeit not exclusively—relating to final goods (Greenaway and Milner 1986).

It follows that those countries that have been more active in off-shoring their activities have experienced both a reduction of the size of their manufacturing potential and a worsening of their trade (im)balances. The rising trade pressure from the emergent world upon industrial countries has to be viewed as the consequence of the production strategies of (some of their) firms. In the perspective suggested by Singh (1977, 1987), this may have also involved in these countries a reduction of efficiency of the manufacturing sector itself, as far as it has both raised their import requirements and lowered their export potential.[13]

In some economic systems such as the Southern European ones this effect has very much been exacerbated by the sheer consequences of the falling down of internal demand, owing to the heavy deflationary policies put in place in order to face sovereign debt crises. Given the general fall in global (international) demand involved by the crisis, this has turned out into a sharp output reduction—in absolute and relative terms—in these countries.

3 Defining Industrial Development

The overall outcome of such changes has been a general blurring of the usual way of grouping countries according to their degree of development. Here the issue is twofold: on the one side, in spite of the still widespread habit of conceiving the world in dichotomous terms, the very emergence of "emerging" countries definitely sets something new between 'North' and 'South'.[14] On the other, the same sharp distinction between "industrial" and "emerging" countries seems in fact not to hold that much, in a context in which important differences seem to emerge *within* these same groups. Put in other words, new differences and similarities may lead to characterise the shape of the industrial world in more articulated way as compared to the past.

This is not simply a matter of classification of economic systems: this is first and foremost a conceptual issue, insofar as it has to do with the definition of what industrial development actually consists of. The point here is that the above-mentioned transformations have determined some changes in the relative distance among all the different countries belonging to the group of manufacturing producers.

[13]According to Singh "an efficient manufacturing sector in an open market context can be defined as one which *(currently as well as potentially) not only satisfies the demand of consumers at home, but is also able to sell enough of its products abroad to pay for the nation's input requirements*. This is, however, subject to the important restriction in that [...] [it] *must be able to achieve these objectives at socially acceptable levels of output, employment, and the exchange rate*" (1977, p. 128, emphasis original).

[14]"The literature is replete with competing terminologies; examples include poor/rich, backward/advanced, underdeveloped/developed, North/South, late comers/pioneers, Third World/First World, and developing/industrialized" (Nielsen 2011, p. 99).

But how do we measure such distance, i.e. the degree of development of different countries?

In the same way as all international institutions typically refer to per capita GDP in order to measure overall economic development, when approaching the task to measure manufacturing development UNIDO refers to per capita manufacturing value added (see Upadhyaya 2013).[15]

Despite its straightforward interpretation, this indicator cannot suffice to the purpose of identifying different paths of industrialisation, because along the continuum of the distribution it is virtually impossible to identify the critical output thresholds according to which the whole population of countries should be divided into relatively homogeneous groups.

The solution proposed by UNIDO in this connection is to group countries according to a mix of alternative rules. These rules do alternatively pay attention to the level of manufacturing value added per capita, by using aprioristic information about the evolution of manufacturing share in GDP to define relevant thresholds, or to the level of GDP per capita or to the national share on world manufacturing value added, by using in both cases ad hoc rules of thumb to define relevant thresholds.

The overall result is the classification of countries into four groups (Industrialized economies, Emerging industrial economies, Other developing countries, Least developed countries).[16] Yet by using very different criteria with reference to different countries, it turns out impossible to determine to what extent countries assigned to the same group are actually homogeneous in terms of their industrial development level.

The view set forth in this paper is that, in order to identify from the ground a taxonomy of manufacturing countries, the challenge is to build metrics which can capture the multidimensional, complex nature of the industrial development process. Such a process—other than the sheer expansion in the national manufacturing sector size—does involve important changes as to the relative manufacturing specialization of the entire economy, the range of manufacturing activities, their technological content (both in the sense of cross-sectoral shifts and of intra-sectoral upgrading processes), and their degree of international competitiveness. All such

[15]The other economic institutions (IMF, World Bank, WTO and the like) that are in charge of ranking individual countries according to their level of development, for operational as well as analytical purposes, shape their definitions on the basis of the characteristics that matter for their specific institutional role (Nielsen 2011). Hence, the IMF pays attention to the developing of financial markets, the United Nations Development Programme to the country's achievements in terms of longevity and education, and so on. For all institutions, anyway, the basic measure to refer to is generally per-capita income. Basically, this means that country grouping boils down to identifying the critical (absolute or relative) thresholds of per-capita income according to which the whole population of countries can be divided, taking into account in some ways also some other variables.

[16]For instance, the group of 'Emerging industrial economies' is identified according to three alternative rules: a level of PPP-adjusted manufacturing value added per capita between 1000 and 2500 US$; a level of PPP-adjusted GDP per capita greater than 10,000 US$; a share in world manufacturing value added greater than 0.5%.

Table 1 Variables used to build the taxonomy

Variable	Mean 2015	S.d.	Mean 2000	S.d.	Source
Manufacturing value added per capita (log, real $)	7.81	1.03	7.51	1.20	UNIDO
Manufacturing value added on GDP (real $)	0.16	0.06	0.15	0.05	UNIDO
Gini index of intra-manufacturing value added (4 digit, real $)	0.62	0.10	0.59	0.11	Oxford Economics
Share of M&High-tech manufacturing value added (real $)	0.49	0.19	0.47	0.17	UNIDO
Manufacturing trade balance, normalized[a] (nominal $)	−0.05	0.15	−0.04	0.18	ComTrade, OEC

[a]Normalization by sum of export and import

transformations are historically determined, knowledge-based and path-dependent, and can be viewed as an evolutionary process in the course of which a given country changes by moving (developing) through different states (phases of growth).[17]

When aiming at classifying countries according to their degree of industrial development, the key issue then becomes to identify the variables which characterise such different states. In empirical terms, capturing these different dimensions is not straightforward, as far as information about some indicators of interest may not exist (maxime for non-OECD countries), or, even when available, it may happen that the way information is collected affects cross-country comparability due to methodological differences. Therefore the choice of the indicators that have to be taken into account needs to balance—in a way which is consistent with the final objective of identifying the most relevant phases of industrial development—the scope of information collected along the different dimensions of interest with the number of countries which can be actually compared.

To this end, the present analysis has been carried out by considering, in addition to the well-established manufacturing value added per capita, four other indicators (Table 1). Each of them is chosen on the basis of specific theoretical premises, and

[17]It is important to stress that this has not to be intended as a mechanical sequence of stages, but simply as the logic to be followed by countries in the course of their development process. The literature on stages of growth has been developed with reference to both economic systems (as in Rostow's (1960) analysis) and firms (mostly in management theory, see McMahon (1998) for all). The approach followed in this paper differs from such a way of interpreting the development process in that it challenges the idea of a unique pattern of development to be applied to all countries, in light of the thought of Gerschenkron (1962), Fuà (1978), Piore and Sabel (1984), Amsden (2001). It may be argued, for example, that laggards must come to terms with global markets that are already structured, and market shares that are already in the hands of incumbents, so that they have to rely upon different strategies with respect to first comers. Or that industrialisation may take place in quite different forms as to the organisation of production activity (hierarchies vs markets), and so on. Even more, as will be clear in the following, in this view countries can move along the development path (passing from one group to another) in both directions—i.e. no achievement can be taken for granted indefinitely.

has to be intended as complementary to each other in characterising an industrial system.

Two of the variables are commonly referred to in the literature on structural change: the well-known share of manufacturing value added in total value added—pioneered by Fisher (1939), Clark (1940), Fuchs (1968), and Kuznets (1973), and recalled in Kaldor's (1966) analysis of increasing returns, hinging on Verdoorn's (1949) approach—and the intra-manufacturing concentration of output—pioneered by Imbs and Wacziarg (2003) and recently also analysed by Romano and Traù (2017). The former is meant to capture the relative direct contribution of manufacturing to GDP, the latter to determine its absolute level of specialisation among the different sub-industries, that is the width of the national supply matrix.

The third variable is represented by the manufacturing trade balance normalized by the level of trade (exports plus imports), taken as a proxy for 'efficiency' of the manufacturing sector in the face of trade openness, in light of Singh's definition (see previous section). The fourth variable is the share of manufacturing valued added produced by medium and high-tech industries, identified according to OECD classification, that provides a measure (although imperfect) of relative intensity of the production technology generated within the domestic manufacturing sector.[18]

Complete information on all these different features of a manufacturing system are retrieved for a sample of 50 countries, from all around the globe. The geographical distribution of the sample is largely rooted in the history of industrialisation—with Europe, North-America and Asia as main players, while Africa, Middle-East and Central Asia remaining largely excluded. It follows that despite being small in absolute terms—as compared to the full list of 144 countries covered by UNIDO statistics—the sample is highly representative of world manufacturing, accounting for around 94% of total value added (at current prices) and 95% of total exports in 2015.[19]

4 Identifying a New Taxonomy

4.1 Which Tiers, and Which Countries Do Belong to Them

The metrics according to which countries are classified is based upon the above-mentioned five variables, and employs them as ingredients of a clustering algorithm.[20] The variables have not been rescaled within a common range of values, as is

[18]Inter-alia, this indicator is referred to in order to measure the degree of attainment of one of the 17 sustainable development goals launched by the United Nations (UNCTAD 2016).

[19]See Appendix 1 for details.

[20]Cluster analysis allows to assign each observation to a specific group on the basis on a multidimensional measure of similarity of their production systems. Two recent applications of cluster analysis to the study of economic development can be found in Zhang and Gao (2015) and Tezanos Vázquez and Sumner (2013). To the best of the knowledge of the authors of this paper, the

often done before applying the clustering algorithm. This is due to the choice of not giving them the same explanatory power (the same weight) in the definition of groups. This way the (log) manufacturing value added per capita, which has a standard deviation eight times larger as compared to the average of the remaining four variables (see again Table 1), is given particular emphasis, and results as the most relevant factor to discriminate different degrees of industrialisation.[21] The number of groups is not defined a priori, but is chosen after having applied the Calinski and Harabasz (1974) optimization method to the results of a Ward's linkage cluster analysis.

As to 2015 (the latest available year), this statistical methodology leads to the identification of five different clusters. However, in order to get groups large enough to make robust statistical analysis, the fifth cluster, composed by only three countries, has been merged to the closer one (i.e. the most similar in terms of manufacturing valued added per capita, so that the final number of clusters has been set to four.[22]

The cluster analysis draws a picture of the different profiles of manufacturing countries which confirms how blurred have become in the course of the Globalisation Age the boundaries between the 'North' and the 'South' of the world, at least when looking at industrial development (Table 2).

First of all, G-7 countries result to be splitted between the groups labeled as '1st tier manufacturers' and '2nd tier manufacturers', with Germany and Japan belonging to the former, while Canada, France, Italy, UK and USA belonging to the latter. The first group comprises also Switzerland, the Scandinavian block plus two Asian tigers, namely Singapore and South Korea (but not Taiwan, which has been included in the second group).[23]

Most of the remaining EU countries are classified as '2nd tier manufacturers', but not all of them: among Eastern European members, only Czechia and Slovakia are included, while Hungary, Poland and Romania are classified as '3rd tier manufacturers', and Bulgaria as one of the '4th tier manufacturers'. Noticeably enough, among Westerns European members Portugal and Greece are classified as '3rd tier manufacturers'. Also BRICS are unevenly distributed among groups, with Brazil,

---

cluster analysis presented in this work is the first attempt to apply it to the study of industrial development.

[21]This choice is consistent with the existing taxonomy proposed by UNIDO (Upadhyaya 2013), while it is in contrast with the logic underlying the construction of the UNIDO Competitive Industrial Performance (CIP) index (see UNIDO various years), which assigns to the manufacturing value added per capita the same weight as to other seven structural indicators.

[22]In particular, Pakistan, India, and Vietnam, with the lowest levels of manufacturing valued added per capita, in the sample, have been identified by the clustering algorithm to form a separated group of manufacturers. The rationale for choosing the group in which to include these countries is provided in the next subsection.

[23]The fourth Asian Tiger, Hong Kong, has been excluded because its development represents a unique case, after China's annexation in 1997.

Table 2 The new taxonomy of manufacturing countries

1st tier manufacturers	2nd tier manufacturers	3rd tier manufacturers	4th tier manufacturers
Austria	Australia	Argentina	Brazil
Denmark	Belgium	Chile	Bulgaria
Finland	Canada	China	Colombia
Germany	Czech Republic	Greece	Ecuador
Japan	France	Hungary	India
Norway	Israel	Malaysia	Indonesia
Singapore	Italy	Mexico	Iran
South Korea	Netherlands	Poland	Morocco
Sweden	New Zealand	Portugal	Pakistan
Switzerland	Slovak Republic	Romania	Philippines
	Spain	Russia	South Africa
	Taiwan	Thailand	Vietnam
	United Kingdom	Turkey	
	United States	Uruguay	
N: 10	N: 14	N: 14	N: 12

Note: Clusters defined according to k-means procedure, using the variables shown in Table 1

China and Russia classified in the 3rd tier, while India and South Africa in the 4th one.

4.2 *Characterisation of the Different Tiers*

But what are the specific characteristics attached to the four groups of manufacturers that have been identified by the cluster analysis?

First of all, there is a clear and strong relation between the sequence of different tiers and the level of manufacturing value added per capita: the industrialisation rates within the 1st tier are always higher than those within the 2nd tier, which in turn are always higher than those within the 3rd tier, which in turn are always higher than those within the 4th tier (Fig. 1). This result suggests *prima facie* an interpretation of the different groups of manufacturers as a sequence of varying degrees of industrial development, from the lowest (corresponding to the 4th tier) to the highest (1st tier).

Yet, such an indicator does not allow on its own to identify which thresholds have to be chosen in order to assign each country to the different groups along the continuum of the distribution. For instance, the distance in the industrialisation rates existing between USA and Norway, showing respectively the highest value within 2nd tier and the lowest within 1st tier manufacturers, is lower than that observed between South Korea and Austria or between Germany and Singapore, despite the fact that these four countries are classified as belonging to the same group.

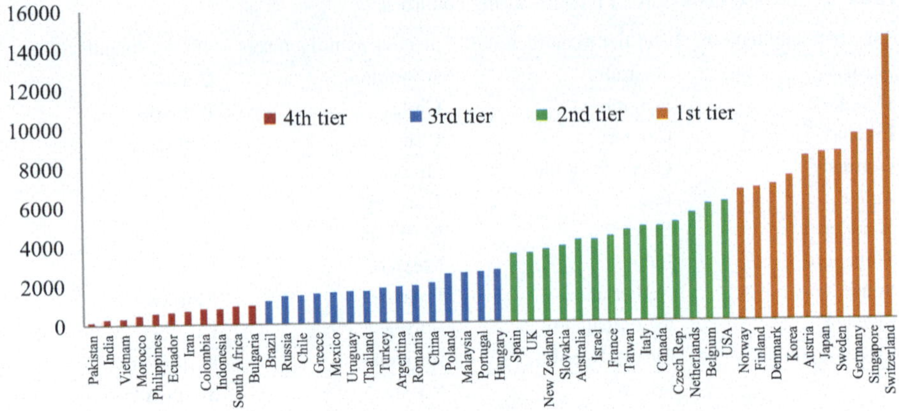

Fig. 1 Industrial development and manufacturing value added per capita (2005, real $)

Table 3 Characterisation of the different tiers of manufacturing countries (2015, mean values)

Variables used in the cluster analysis:	4th tier	3rd tier	2nd tier	1st tier
Manufacturing value added per capita (log)	6.29	7.52	8.41	9.04
Gini index of manufacturing VA	0.71	0.61	0.58	0.58
Share of Medium, High-Tech manufacturing VA	0.31	0.33	0.43	0.56
Manufacturing trade balance normalized	−0.22	−0.04	−0.05	0.06
Share of manufacturing VA on GDP	0.15	0.17	0.13	0.17
		Difference		
		4th–3rd tier	3rd–2nd tier	2nd–1st tier
Manufacturing value added per capita (log)		**−1.23**∗∗∗	**−0.89**∗∗ ∗	**−0.62**∗∗∗
Gini index of manufacturing VA		**0.10**∗∗	0.04	−0.01
Share of Medium, High-Tech manufacturing VA		−0.02	**−0.11**∗∗	**−0.13**∗∗
Manufacturing trade balance normalized		**−0.18**∗∗	0.01	**−0.11**∗∗
Share of manufacturing VA on GDP		−0.01	0.04	**−0.04**∗

Note: Manufacturer tiers defined according to Ward's linkage procedure
∗: p-value < 0.10, ∗∗: p-value < 0.05, ∗∗∗: p-value < 0.01

On the other hand, the four groups also differ from each other in terms of the values assumed by the other four indicators included in the clustering algorithm. In particular, each indicator plays a role to discriminate between different pairs of groups, varying depending on the chosen comparison (Table 3).[24]

[24]The cross-cluster analysis that follows is not driven by the presence of outliers, as similar conclusions can be drawn by comparing median instead of the mean values for each variable in the different groups.

First, as compared to 3rd tier ones, countries classified in the 4th tier show on average a significantly lower degree of diversification (higher Gini index) of manufacturing activities, and a worse positioning in international markets, as reflected by a five times larger trade deficit. No systematic differences emerge instead between the two groups in terms of the relative contribution of manufacturing to GDP and their technological intensity. In conceptual terms, this means that the widening of the supply matrix and the contemporary improvement of the trade balance are the two conditions that go along with the transition from a low to a medium-low degree of industrialisation rate.

Second, 3rd tier countries are in turn significantly less specialized in medium and high-tech industries as compared to 2nd tier ones, while no systematic differences between the two groups emerge in terms of either sectoral concentration of manufacturing value added (both relative to GDP and within manufacturing) or trade balance. Put in other words, the foremost indicator which characterizes the transition from a medium-low to a medium-high degree of industrialisation rate seems to be a higher capacity to generate production technology domestically.

Third, in comparison to 1st tier countries, 2nd tier ones show a systematically lower degree of specialization in medium and high-tech industries, a worse positioning in international markets (the 1st tier is the only cluster with a positive manufacturing trade balance on average), and also a lower share of GDP accounted for by domestic manufacturing. This means that the moving from a medium-high to a high degree of industrialisation rate results in a further increase of the capacity to generate production technology domestically, a significantly better trade balance, and a higher industrial specialization of the domestic economy.

4.3 New and Old Taxonomy in Comparison

It is worthwhile to compare the results of the cluster analysis carried out on the basis of the above-mentioned indicators with those provided by the UNIDO classification of these same countries. Data reveal in this connection some important differences (Table 4).

According to the UNIDO classification the entire group of 1st tier and of 2nd tier, as well as a bunch of 3rd tier countries, namely Malaysia, Portugal and Russia, are assumed to belong to the same stage of development—and are coherently classified as 'Industrialized economies'. The rest of 3rd tier manufacturers and half of 4th ones are instead grouped together as 'Emerging industrial economies', while remaining '4th tier manufacturers' are classified as 'Other developing countries'.

On the whole, the UNIDO classification, with some important exceptions (mainly within the group of less industrialised economies), in contrast with the taxonomy proposed in this work, tends to replicate the well-established dichotomy between 'North' and 'South' of the world.

Moreover, the UNIDO classification draws a picture in which the majority of manufacturing countries (52% of the sample) is assumed to be at the latest stage of

Table 4 Comparing with UNIDO taxonomy of industrial development in 2015

Industrialized economies		Emerging industrial economies		Other developing economies	
1st tier	Austria	3rd tier	Argentina	4th tier	Iran
"	Denmark	"	Brazil	"	Morocco
"	Finland	"	Chile	"	Pakistan
"	Germany	"	China	"	Philippines
"	Japan	"	Colombia	"	Vietnam
"	Norway	"	Greece		
"	Singapore	"	Hungary		
"	South Korea	"	Mexico		
"	Sweden	"	Poland		
"	Switzerland	"	Romania		
2nd tier	Australia	"	Thailand		
"	Belgium	"	Turkey		
"	Canada	"	Uruguay		
"	Czech Republic	4th tier	Bulgaria		
"	France	"	Ecuador		
"	Israel	"	India		
"	Italy	"	Indonesia		
"	Netherlands	"	South Africa		
"	New Zealand				
"	Slovak Republic				
"	United Kingdom				
"	Spain				
"	Taiwan				
"	United States				
3rd tier	Malaysia				
"	Portugal				
"	Russia				
	N: 27		N: 18	N: 5	

Note: UNIDO classification as from Upadhyaya (2013)

industrial development, while, according to the analysis carried out in this paper, the most advanced manufacturing powerhouses represent only a minority (20%) of the sample.

4.4 *Evolution of the Taxonomy Over Time*

One of the benefits of approaching the definition of a new taxonomy relying on observed data is that, by construction, the belonging of each country to a specific cluster as well as the number of clusters are not defined a priori, but can and do

Table 5 The taxonomy of manufacturing countries in 2001

1st tier	2nd tier	3rd tier	4th tier	5th tier
Austria	Australia	Czech Republic	Argentina	Bulgaria
Belgium	France	Greece	Brazil	China
Canada	Israel	Hungary	Chile	Colombia
Denmark	New Zealand	Malaysia	Poland	Ecuador
Finland	South Korea	Mexico	Romania	India
Germany	Spain	Portugal	Russia	Indonesia
Italy	Taiwan		South Africa	Iran
Japan	United Kingdom		Thailand	Morocco
Netherlands			Turkey	Pakistan
Norway			Slovak Republic	Philippines
Singapore			Uruguay	Vietnam
Sweden				
Switzerland				
United States				
N: 14	N: 8	N: 6	N: 11	N: 11

Note: Clusters defined according to Ward's linkage procedure, using the variables shown in Table 1. Colors are assigned to different tiers of manufacturers according to the 2015 taxonomy, as to Table 2.

change in response to the different paths of industrial development observed over time.

In this respect, as explained in Sect. 2, at the dawn of the new millennium major changes at the global level had already occurred, mostly due to the 'unfreezing' of large economic systems that still in early 90s were sealed within a parallel world, outside of the area of market exchanges. Therefore, the cluster analysis should be able to reflect such a fragmentation of global manufacturing, before the overall reshaping brought about by the Globalisation Age.

Consistently with that, the analysis reveals that in 2001 five different groups of manufacturers could be detected, i.e. one more as compared to the present (Table 5). In particular, a mix of the least developed Western European economies (Greece and Portugal), of the most developed Eastern European ones (Czechia and Hungary), together with Malaysia and Mexico formed a relatively homogeneous group in the middle of the distribution of manufacturing value added per capita.[25]

All these countries, as already shown in Table 2, were then absorbed in the current group of 3rd tier manufactures, with the exception of Czechia that was able to join the 2nd tier. A significant reshuffling has also occurred within the group of most advanced manufacturing countries: Belgium, Canada, Italy, Netherlands and the United States have exited the group of 1st tier, joining the remaining old

[25]As for the 2015 taxonomy, also in this case there is an unambiguous relation between the sequence of different tiers and the level of manufacturing value added per capita. Moreover, also in this case, a further (sixth) cluster including the three least developed manufacturing countries (Pakistan, India, and Vietnam) has been isolated by the hierarchical clustering algorithm. Just as for 2015, this group has then been merged with the adjacent one.

industrialised economies in the 2nd tier. South Korea, on the opposite, has become one of the leading manufacturing countries in the world, joining the 1st tier.

Within the groups of the least developed manufacturing countries, changes have characterised especially the relative position of BRICs: while Brazil and South Africa have moved towards the bottom, joining the current group of 4th tier manufacturers, China, on the opposite, has been able to climb from the bottom, joining the current group of 3rd tiers.

Of course, there is also significant heterogeneity in the underlying dynamic performance of different countries, even for those that have experienced the same transition from one tier to another. For instance, the shift of Italy and Canada into the current 2nd tier is associated with a reduction in their manufacturing real value added per capita (−15.0 and −21.3% respectively along the 2001–2015 period), while the same movement of Belgium, Netherlands and the United States is associated with positive growth of the same indicator (4.1, 4.4 and 10.8% respectively), although at a slower pace as compared to the remaining 1st tier manufacturers (16.7%). At the same time, although both China and Russia are currently classified in the same tier, their growth in the last 15 years has been remarkably different: the former has more than doubled its industrialisation rate (+134.5%) while the latter has expanded at a speed (+36.1%) which has been below the average of remaining countries currently belonging to the 3rd tier (+41.6%). A negative record has instead characterised Greece and Portugal, the only two countries currently belonging to the 3rd tiers with negative manufacturing real value added growth between 2001 and 2015 (−38.7 and −8.9%, respectively).

Overall, by applying the same methodology described in the previous section to year 2001, it is possible to appreciate how the impressive acceleration of the globalisation process in the last 15 years, as well as the tremendous and asymmetric effects of the Great Recession, have reshaped the geography of world manufacturing. We can expect new movements across tiers to occur also in thc future and new groups to possibly emerge. The methodology proposed in this paper is flexible enough to (hopefully) being able to accommodate for such changes.

5 Institutional Features

Institutions matter for development. They define the rules of the game according to which economic actors interact with each other and the scope and width of policy intervention in the economy (North 1990; Acemoglu and Robinson 2008), influencing investment decisions, the accumulation of knowledge, and the whole organization of production. Indeed, as stressed, inter alia, by Amsden (2001), Acemoglu et al. (2001, 2002), Chang (2002, 2003), Cimoli et al. (2009), and Romano and Traù (2014), the different roles interpreted by public institutions in order to support industrial development in the course of the twentieth century have been at the heart of the divergence in fortune observed among emerging economies. Significant institutional and policy heterogeneity also has characterized the evolution of

Western economies, despite the relatively higher degree of similarity (and harmonization) of their legal and regulatory frameworks.[26]

But institutions are not chosen once for all: they respond in the course of development to new societal and economic challenges which gradually do emerge, implying that also policies tend to (and should) be contingent on each phase of development (Hausmann et al. 2005). This is in line with Lee and Kim (2009), showing how different institutional arrangements and policies have mattered for economic growth in the last decades, but with varying intensity depending on the degree of development of countries.

Given this premise, one may ask whether—and to what extent—the different phases of industrialisation described in the previous section also mirror cross-country differences as to the institutional side. Cross-country analysis does not allow to make any causal inference about the relation between changes in the institutional setting and industrial development; nevertheless, it can be useful for the purpose of detecting which institutional features the different tiers of manufacturers are characterized by, so as to verify the existence of something like an 'institutional progression' (as suggested by Meisel and Ould Aoudia 2008) when moving from one phase of development to the next.

To this end, the present analysis has been enriched by looking at the Institutional Profiles Database built jointly by the Agence Française de Développement (AFD), the French Direction Générale du Trésor (DG Trésor) and CEPII (2016).[27] In particular, information about 46 institutional indicators for the 50 countries under consideration has been retrieved from the original database and compared across groups of manufacturers, to see which of them show systematic differences when moving from a tier to the next.[28] Such indicators, measured in ordinal (Likert) scale (from 0 to 4), are classified in the original dataset according to the following six functions: Functioning of public system, Free operation of markets, Security of transactions, Market regulations, Openness (to foreign exchanges) and Coordination of stakeholders, Strategic vision and innovation.[29]

The first five functions basically overlap the 'classical' good governance indicators used by the World Bank, which in turn reflect the old-fashioned idea behind the Washington Consensus of a desirable State intervention mainly (if not exclusively) limited to making domestic markets work better (Stiglitz 1999). The sixth group, instead, echoes a novel approach to industrial policy, where the State is supposed to "perform a strategic and coordinating role in the productive sphere beyond simply ensuring property rights, contract enforcement, and macroeconomic stability" (Rodrik 2004, p. 3), affecting the sectoral composition of the economy and the transition towards new technological paradigms (Stiglitz 2015).

---

[26]See, for instance, Mazzucato (2011) with reference to public policies oriented towards technological innovation.

[27]To download the database and read related documentation: www.cepii.fr/institutions/EN/ipd.asp

[28]See Appendix 2 for the list of indicators.

[29]See Kaufmann et al. (2010) for details.

The cross-tier comparison of the indicators appears informative in different respects (Table 6). It shows that important institutional differences do characterise each transition. This involves that challenges and priorities for industrial policy need to vary in intensity along the different phases of industrialisation. In particular, the most important discontinuity characterises the advancement from the 3rd to the 2nd tier of manufacturing countries. This is true considering both the number of indicators for which (statistically significant) differences have been detected and the range of institutional functions involved. Therefore, results are consistent with the view according to which the institutional progression across the different phases of industrialisation does not occur smoothly but requires at some point of industrial development—in particular when moving from a medium-low to a medium-high industrialisation rate—a radical change in the functioning of the public system, in the legal and regulatory framework surrounding market transactions, in the openness to international markets and in the degree of strategic collaboration and coordination with the private sector.

Moreover, the analysis shows that a significant portion of the institutional differences refer to the active role played by the State in curbing structural change through active support to technological innovation and coordination of public-private initiatives. This is true in all transitions, but it is particularly relevant within the group of the most advanced economies, where such institutional differences account for half of the total. In particular, 1st tier manufacturers are, on average, on a higher rank as compared to 2nd tier manufactures in terms of State long-term vision, public support for innovation and adaptation of training supply to business needs. Conversely, the large majority of standard market-friendly institutional arrangements, including those related to labour market rigidity or competition barriers, are systematically different only between 3rd and 2nd tier manufacturers.

6 Concluding Remarks

The analysis proposed in this work has shown that industrial development is not only a matter of expanding the scale of manufacturing production. Changes in the degree of industrialisation go along with important differences in the manufacturing output share, in the sectoral composition of the production system, as well as in its external competitiveness.

Having identified different groups (tiers) of countries, corresponding to various degrees of industrial development, however, does not imply that countries are bound to follow, when developing, a mechanical path leading from one tier to another. Countries do not follow a pattern of development which is given once and for all, and the same characteristics that nowadays discriminate between different tiers of manufacturers could lose significance, change sign, or being replaced by new ones tomorrow, should changes in the prevailing technological and institutional paradigms occur, as happened in the past.

Does this imply that nothing can be said in order to inform policy-making on how to sustain industrial development? Giving precise policy prescriptions or

Table 6 Institutional differences across tiers of manufacturing countries (2015)

Functioning of public system	Free operation of markets	Security of transactions	Market regulations	Openness	Coordination of stakeholders, strategic vision, innovation
Moving from 4th to3rd tier					
+ capacity of sectoral reforms		– security of private contracts	+ importance of laige-scale distribution		+ public support for innovation
– tax exemptions to economic sectors					
Moving from 3rdto 2nd tier					
+ capacity of sectoral reforms	– share of state-owned companies	+ security of private contracts	– barriers to competition	– obstacles to trade liberalisation	+ coordination in the public sphere
+ transparency	+ efficiency of state-owned companies	+ trade justice	+ competition regulation	– obstacles to capital liberalisation	+ priority to development and growth
+ efficiency of tax system and justice	– state-owned banks	+ insolvency law			+ long-term sectoral strategies
+ freedom to establish and operate organisations	+ mobility of workers	– termination of contracts by State			+ technological environment for firms
	+ ease of starting a business	+ information on firms			+ adaptation of training and schooling to business needs
	– pricing control	+ respect for intellectual property			
Moving from 2nd to 1st tier					
+ efficiency of tax system		+ respect for intellectual property			+ state long-term vision
+ quality of the public policy making					+ public support for innovation
					+ adaptation of the training supply to business needs

Note: Table reports only variables for which the difference between groups is statistically significant with a p-value < 0.05

recommendations goes well beyond the scope of this work, because this would require an in-depth analysis of the temporal evolution of both the institutional setting and the manufacturing system in each country. However, two general lessons for policy can be learned from the previous results.

First, the institutional progression that characterizes the actual transition from one tier of manufacturers to the next confirms the idea that enforcing the same development strategy in different phases of development is unlikely to prove effective in reaching the policy objective. In Hausman and co-authors' words, "[g]overnments face administrative and political limitations, and their policy-making capital is better deployed in alleviating binding constraints than in going after too many targets all at once. So growth strategies require a sense of priority" (Hausmann et al. 2005, p. 2). In this respect, excessive emphasis on making institutions more market-oriented could be misguiding for countries that are either very low or very high in terms of industrialisation, while it is appropriate for countries being in the middle range of industrialisation.

Second, results suggest that for a correct understanding of the policy challenges that go along with industrial development, a more comprehensive look at institutions, taking into account the role of the State in actively promoting and coordinating private investments, is necessary. However, the functions and scope of State involvement in the economy is not the same in all phases of development, but it should be fine-tuned to the actual needs of industry.

Appendix 1 World shares of manufacturing countries in the sample (2005, nominal $)

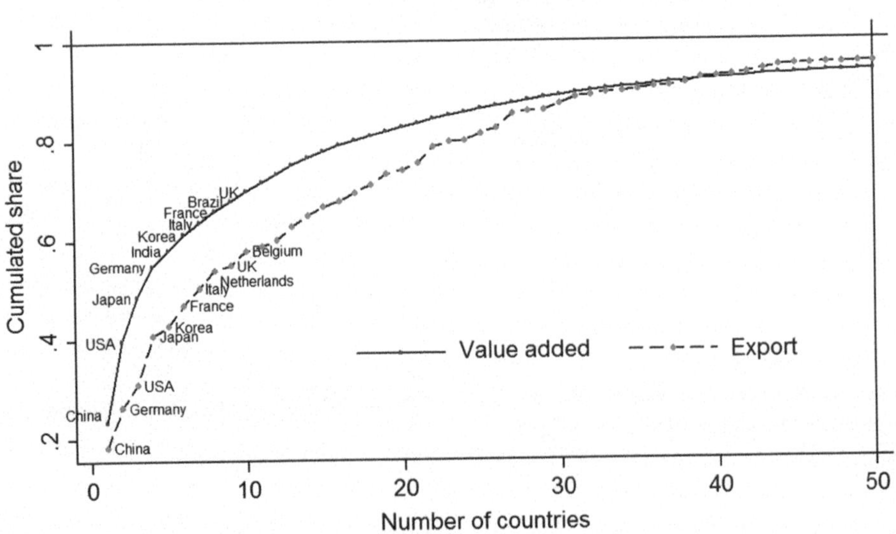

Appendix 2 List of selected institutional indicators

Functioning of public administrations	Free operation of markets	Security of transactions and contracts
Transparency of economic policy	Significance of public companies to the economy	Security of private contracts
Efficiency of the tax administration	Accounts of State-owned and partly State-owned firms	Trade justice
Transparency in public procurement	Weight of State-owned banks	Insolvency law
Functioning of the justice system	Labour market rigidity	Termination of contracts by the State
Influence of economic stakeholders	Mobility of workers	Respect for intellectual property
Freedom to establish organisations	Pricing control	Employment contract protection
Ease of starting a business		
Quality of the public policy making process		
Support for emerging dynamic sectors		
Consideration of the public interest in State-business relation		
Political authorities decision-making autonomy		
Quality of public services (provided by the public sector)		
Market regulations	Openness	Coordination of stakeholders, strategic vision, innovation
Competition: barriers to market entry	Trade liberalisation	Capacity for State reform
Importance of large-scale distribution	Obstacles to trade liberalisation	Capacity for sectoral reform
Competition regulation	Obstacles to financial liberalisation	Tax exemptions
Scale of public ownership	Importance of joint ventures in the economy	Public-private cooperation
Information on the capital held by firms	Opening up of the financial system	Coordination in the public sphere
		Long-term vision
		Long-term sectoral strategies
		Spaces for reflection on the major national issues
		Priority of the elite in relation to development and growth
		Technological environment of firms

(continued)

Functioning of public administrations	Free operation of markets	Security of transactions and contracts
		PubEc support for innovation
		Venture capital
		Adaptation of the training supply to business needs
		Adaptation of the higher education system to business needs

Source of selected indicators: Institutional Profile Database

References

Acemoglu, D., & Robinson, J. A. (2008). *The role of institutions in growth and development.* Commission on Growth and Development Working Paper 10.

Acemoglu, D., Johson, S., & Robinson, J. A. (2001). The colonial origins of comparative development: An empirical investigation. *American Economics Review, 91*(5), 1369–1401.

Acemoglu, D., Johson, S., & Robinson, J. A. (2002). Reversal of fortune: Geography and institutions in the making of the modern world income distribution. *The Quarterly Journal of Economics, 117*(4), 1231–1294.

Amsden, A. (2001). *The rise of the rest: Challenges to the west from late-industrializing economies.* New York: Oxford University Press.

Baldwin, R. (2006). *Globalization: The great unbundling(s).* Prime Minister's Office, Economic Council of Finland.

Baldwin, R. (2012). *Global supply chains: Why they emerged, why they matter, and where are they going.* CTEI Papers, 2012–13. Geneva: The Graduate Institute, Centre for Trade and Economic Integration.

Baldwin, R. (2014). Trade and industrialization after globalization's 2nd unbundling: How building and joining a supply chain are different and why it matters. In R. C. Feenstra & A. M. Taylor (Eds.), *Globalization in an age of crisis: Multilateral economic cooperation in the twenty-first century.* Chicago: Chicago University Press.

Calinski, T., & Harabasz, J. (1974). A dendrite method for cluster analysis. *Communications in Statistics, 3*(1), 1–27.

Cattaneo, O., Gereffi, G., & Staritz, C. (Eds.). (2010). *Global value chains in a post-crisis world: A development perspective.* Washington: The World Bank.

Chang, H. J. (2002). *Kicking away the ladder: Development strategy in historical perspective.* London: Anthem Press.

Chang, H. J. (2003). Introduction. In H. J. Chang (Ed.), *Rethinking development economics.* London: Anthem Press.

Cimoli, M., Dosi, G., & Stiglitz, J. E. (Eds.). (2009). *Industrial policy and development.* New York: Oxford University Press.

Clark, C. (1940). *The conditions of economic progress.* London: Macmillan.

Feenstra, R. C. (1998). Integration of trade and disintegration of production in the global economy. *Journal of Economic Perspectives, 12*(4), 31–50.

Fisher, A. (1939). Production: Primary, secondary and tertiary. *Economic Record, 15*(1), 24–38.

Freeman, R. B. (2007). The great doubling: The challenge of the new global labor market. In J. Edwards, M. Crain, & A. L. Kalleberg (Eds.), *Ending poverty in America. How to restore the American dream.* New York: The New Press.

Fuà, G. (1978). Lagged development and economic dualism. *Banca Nazionale del Lavoro Quarterly Review, 125*, 123–134.
Fuchs, V. R. (1968). *The service economy*. NBER General Series n. 87.
Gereffi, G. (2014). Global value chains in a post-Washington consensus world. *Review of International Political Economy, 21*(1), 9–37.
Gereffi, G., Humphrey, J., & Sturgeon, T. (2005). The governance of global value chains. *Review of International Political Economy, 12*(1), 78–104.
Gerschenkron, A. (1962). *Economic backwardness in historical perspective*. Cambridge, MA: Harvard University Press.
Greenaway, D., & Milner, C. (1986). *The economics of intra-industry trade*. Oxford and New York: Blackwell.
Hausmann, R., Rodrik, D., & Velasco, A. (2005). *Growth diagnostics*. Cambridge, MA: John F. Kennedy School of Government, Harvard University.
Hirschman, A. (1958). *The strategy of economic development*. New Haven: Yale University Press.
Hirschman, A. O. (1981). A generalized linkage approach to development, with special reference to staples. In A. O. Hirschman (Ed.), *Essays in trespassing*. Cambridge: Cambridge University Press.
Imbs, J., & Wacziarg, R. (2003). Stages of diversification. *American Economic Review, 93*(1), 63–86.
Kaldor, N. (1966). *Causes of the slow rate of economic growth in the United Kingdom*. Cambridge: Cambridge University Press.
Kaufmann, D., Kraay, A., & Mastruzzi, M. (2010). *The worldwide governance indicators: Methodology and analytical issues*. Policy Research Working Paper WPS 54. World Bank.
Kuznets, S. (1973). Modern economic growth: Findings and reflections. *American Economic Review, 63*(3), 247–258.
Lee, K., & Kim, B. Y. (2009). Both institutions and policies matter but differently for different income groups of countries: Determinants of long-run economic growth revisited. *World Development, 37*(3), 533–549.
Lewis, W. A. (1954). Economic Development with Unlimited Supplies of Labour. *The Manchester School, 22*(2), 139–191.
Lin, J. Y. (2009). *Economic development and transition. Thought, strategy, and viability*. New York: Cambridge University Press.
Mazzucato, M. (2011). *The entrepreneurial state*. London: Demos.
McMahon, R. G. P. (1998). Stage models of SME growth reconsidered. *Small Enterprise Research, 6*(2), 20–35.
Meisel, N., & Ould Aoudia, J. (2008). *Is good governance a good development strategy?* Agence Française de Développement Working Paper 58.
Nielsen, L. (2011). *Classification of countries based on their level of development: How it is done and how it could be done*. IMF Working Paper, 11/31.
Nolan, P., Zhang, J., & Liu, C. (2008). The global business revolution, the cascade effect, and the challenge for firms from developing countries. *Cambridge Journal of Economics, 32*(1), 29–47.
North, D. (1990). *Institutions, institutional change and economic performance*. Cambridge: Cambridge University Press.
Piore, M. J., & Sabel, C. F. (1984). *The second industrial divide*. New York: Basic Books.
Rodrik, D. (2004). *Industrial policy for the twenty-first century*. KSG Working Paper RWP04-047.
Romano, L., & Traù, F. (2014). Il ruolo delle istituzioni nello sviluppo manifatturiero del mondo emergente. Tre modelli di intervento pubblico negli anni successivi al secondo dopoguerra. *Rivista di Storia Economica, 30*(2), 121–159.
Romano, L., & Traù, F. (2017). The nature of industrial development and the speed of structural change. *Structural Change and Economic Dynamics, 42*, 26–37.
Rostow, W. W. (1960). *The stages of economic growth*. Cambridge: Cambridge University Press.
Singh, A. (1977). UK industry and the world economy: A case of de-industrialisation? *Cambridge Journal of Economics, 1*(2), 113–136.

Singh, A. (1987). Manufacturing and de-industrialisation, entry. In J. Eatwell, M. Milgate, & P. Newman (Eds.), *The new Palgrave*. London: Macmillan.
Sinn, H. W. (2006). The pathological export boom and the bazaar effect: How to solve the German puzzle. *The World Economy, 29*(9), 1157–1175.
Stiglitz, J. E. (1999). More instruments and broader goals: Moving toward the post-Washington consensus. In G. Kochendorfer-Lucius & B. Pleskovic (Eds.), *Development issues in the 21st century*. Berlin: German Foundation for International Development.
Stiglitz, J. E. (2015). *Industrial policy, learning, and development*. WIDER Working Paper 2015/149.
Sturgeon, T. J. (2008). *From commodity chains to value chains: Interdisciplinary theory building in an age of globalization*. Industry Studies Association Working Papers, WP 2008-02.
Tezanos Vázquez, S., & Sumner, A. (2013). Revisiting the meaning of development: A multidimensional taxonomy of developing countries. *Journal of Development Studies, 49*(12), 1728–1745.
Traù, F. (2003). *Structural macroeconomic change and the size pattern of manufacturing firms*. Basingstoke: Palgrave Macmillan.
Traù, F. (2016). La globalizzazione e lo sviluppo industriale mondiale. *Rivista di Politica Economica, 105*(10–12), 353–408.
UNCTAD. (2016). *Development and globalization: Facts and figures*. Geneva: United Nations.
UNIDO. (2009). *Industrial development report. Breaking in and moving up: New industrial challenges for the bottom billion and the middle-income countries*. Vienna: United Nations.
UNIDO (various years): *Competitive industrial performance report*. United Nations. Vienna.
Upadhyaya, S. (2013). *Country grouping in UNIDO statistics*. UNIDO Working Paper n. 1.
Verdoorn, P. J. (1949). Fattori che regolano lo sviluppo della produttività del lavoro. *L'Industria, 1*, 45–53.
WTO, IDE-JETRO. (2011). *Trade patterns and global value chains in East Asia: From trade in goods to trade in tasks*.
Zhang, Z., & Gao, Y. (2015). *Emerging market heterogeneity: Insights from cluster and taxonomy analysis*. IMF Working Paper 15/155.

Spatial-Sectoral Skill Polarization: Is South of Italy Not Lost?

Martino Lo Cascio and Massimo Bagarani

Abstract A new paradigm seems to emerge in the international division of labor in Europe and in the most advanced production systems all over the world: on the one hand, labor patterns in knowledge-intensive sectors show different trajectories compared with the observed performance of employment in other sectors.

In this framework, employment in Italy shows a similar behavior, but at a slower pace of convergence to this new paradigm, at least with respect to its main competitors (Germany, France, Great Britain and, in perspective, Spain).

Differences emerge among its regional macro-areas: Southern Italy is diverging from the rest of the overall Italian economy. This area, despite the dimension of the current gap with the rest of Italy, can traditionally count on a higher dynamism in the "creative destruction" process necessary to respond to external threats than of the Center and the North of Italy. This may give room for a cautious optimism about the possibilities of the South to resume an important role in the long-term growth of the country.

In this paper, using constrained logistic functions, we try modeling diverse forms of labor substitution, comparing the dynamics of the structural gaps between Southern Italy and the rest of the country and providing an answer to the question: "Is the South of Italy definitively lost?"

Keywords Italy · Technology · Employment · Manufacturing · EUROSTAT

JEL D01 · E00 · J24 · R10

M. L. Cascio (✉)
University of Rome "Tor Vergata", Rome, Italy
e-mail: martino.locascio@uniroma2.it

M. Bagarani
"Guglielmo Marconi" University, Rome, Italy
e-mail: m.bagarani@unimarconi.it

L. Paganetto (ed.), *Capitalism, Global Change and Sustainable Development*, Springer Proceedings in Business and Economics,
https://doi.org/10.1007/978-3-030-46143-0_13

1 Introduction

This contribute moves from the current literature on the development theory as surveyed by Jones and Romer (Jones and Romer 2010; Romer 1986), coupled with our previous approaches on *complexity theory based* toolbox (Brian Arthur 1989; Brian 2013), on incoming labor product society and total factor productivity transfers within and between economic areas.

Jones and Romer updated the six Kaldor "stylized" facts in a new six facts (Kaldor 1961):

1. Increases in the extent of the market;
2. Accelerating growth;
3. Variation in modern growth rare;
4. Large income and TFP differences;
5. Increases in human capital per worker;
6. Long-run stability of relative wages.

The basic features of Romer approach are the introduction of:

1. The concept of "Ideas" as a development factor even in the case they are not rival (and the linked increased extent of the market);
2. The interaction between Ideas, Institutions, Population and Human Capital as an essential loop for a growth model;
3. The extraordinary rise in the extent of the market associated with globalization and urbanization, that, in the Romer words, "are really understood as reflecting the defining characteristic of ideas, their nonrivarly";
4. The emphasis in modern growth theory on institutions and human capital, ("human capital per worker is rising rapidly and this occurs despite a no systematic trend in the wage premium associated with education").

Increased flows of goods, ideas, finance, and people—via globalization as well as urbanization—have increased the extent of the market for all workers and consumers.

The loop between not rival ideas-innovation-population-not rival ideas, coupled with the increasing importance of human capital, reveal important complementarities among the key endogenous variables, suggesting to put population/labor as a support variable for:

- *A*, set of ideas;
- *L*, individuals who can supply labor in an aggregate development model, whose argument include
- *X*, a vector of the more traditional rival physical goods and services.

For Romer, at least until now, it is impossible to endogenize *Institutions*, or politics. For us is possible to think to a novel Human Economics, but not to an economic model of the World, even in the future.

However, the two of us have begun to undermine the *first absurd habit* of a single nation based equation and the consequent identification of the six above mentioned facts, in order to deal with the convergence/divergence between non-closed economies.

In "Incoming Labor-Product Society and EU Regional Policy" (Lo Cascio and Bagarani 2018, but also Lo Cascio and Bagarani 2012), we introduced "myopic steady-state convergence model", in the complexity meso-economics framework, where:

- The steady state condition, assumed as continuously shifting, represents the result and a new impulse for a *platea* of adapting decision makers (this is why we call ex-post and myopic this kind of model);
- A chained Total Factor Productivity Transfer (TFPT) is included as a relevant factor in explaining the regional convergence/divergence process;
- A procedure linking the estimated fixed effects to institutional regional features was included;
- The labor- product share, among the arguments, are included.

Regarding last one variable, in that work, we proposed the category of labor-product as opposed to labor as a factor of production. Coexisting forms of labor in societies range from an upper class, the labor-product class, to different types of post-fordist labor, where the intensity of technology is declining (Lo Cascio 2019).

The introduction in the model of the last new variable improved substantially the statistics of the estimates and the economic meaning of the results, showing a positive effect of labor-product share on growth and a negative on the convergence mechanism. The last one result seems to be relevant, asking for an adapting policy machine to the evolving conditions of knowledge in time and space (different situations ask for a unique strategy, but different operating rules among the territories).

Some further information, compared to previous studies, emerge from the last one approach even if still remain doubts on specification and estimates. No perspective toward the inclusion of the *Institutions* as endogenous is envisaged.

The new century globalization in trade and finance—both between euro area and other world economic areas and within euro zone—quickens the pace of "creative destruction" and thereby speeds the flow of technology across European countries in the long term.

Furthermore, Jones and Romer's theoretical survey and proposals coupled with the complexity based meso and macro modeling are a way to understand, clarify and explain the core of the logistics of European models as introduced in "Great European Crisis: Shift or Turning Point in Job Creation from Job Destruction" (Lo Cascio and Bagarani 2019), i.e.:

1. accepting the *second absurd habit* of standard macroeconomics, *the representative agent*;
2. trying to stylize the production function with labor-product arguments both on the right and on the left side (with exogenous institutions). Being solutions included

in a range between an (impossible) upper limit as it is in the stable exponential growth (the Earth is finite) and a series of lower bounds linked to a continuously updating and shifting of subsistence levels in consumption and production, in a sort of new Malthusian or Ramsey misery and bliss perspective.

Following the previous study on European logistic models (Lo Cascio and Bagarani 2019), we adopt a further development of the naïve model of diverse forms of labor substitution [constrained logistic functions, as that, for primary energy sources, of Marchetti and Nakicenovic (1979)].

In these naïve logistic models, population and total labor levels are exogenous, like total demand/supply for energy in IIASA studies. So that, the logic of the approach, something like technology and society in other Marchetti contributions (Marchetti 2003), is allowing us to explore the structural substitutions in time between labor knowledge clusters, their inertia and varying momentum as emerging from previous long term history. On statistical ground, we use estimates that minimize loss of information functions, over the time, of the log odds ratios of labor for different levels of training and technological qualification, starting from a MLE procedure that assures the satisfaction over time of the constrain to one of the predicted cluster's shares.

2 The Dataset from EUROSTAT Database

The analysis utilized data from the EUROSTAT database and, in particular, statistics on "Employment in technology and knowledge-intensive sectors by NUTS 2 regions" for the period 1994–2018, in EUROSTAT "High-tech industries and knowledge-intensive services data".

The EUROSTAT source proposes a reading of employment data through an innovative typological aggregation of sectors, based not only on the type of product, but also on the degree of embedded technology in manufacturing and services sectors.

The data provided are therefore related to employment in production sectors differentiated by product and by levels of technology intensity:

A_B	Agriculture, forestry and fishing; mining and quarrying
HTC	High-technology sectors (high-technology manufacturing and knowledge-intensive high-technology services)
C	Manufacturing
C_HTC_MH	High and medium high-technology manufacturing
C_HTC_M	Medium high-technology manufacturing
C_HTC	High-technology manufacturing
C_LTC_LM	Low and medium low-technology manufacturing
C_LTC_M	Medium low-technology manufacturing
C_LTC	Low-technology manufacturing

(continued)

D-F	Electricity, gas, steam and air conditioning supply; water supply and construction
G-U	Services
G_I_T	Wholesale and retail trade; accommodation and food service activities; activities of households as employers
H49-H52_N79	Land transport, transport via pipelines, water transport, air transport, warehousing and support activities for transportation; travel agency, tour operators
KIS	Total knowledge-intensive services
KIS_HTC	Knowledge-intensive high-technology services
KIS_MKT_OTH	Knowledge-intensive market services (except financial intermediation and high-technology services)
KIS_OTH	Other knowledge-intensive services
LKIS	Total less knowledge-intensive services
LKIS_MKT	Less knowledge-intensive market services
LKIS_OTH	Other less knowledge-intensive services
J	Information and communication
K_L	Financial and insurance activities; real estate activities
K	Financial and insurance activities
M	Professional, scientific and technical activities
N	Administrative and support service activities
O_U	Public administration; activities of extraterritorial organizations and bodies
P	Education
Q	Human health and social work activities
R	Arts, entertainment and recreation
S	Other service activities
OTH	Other NACE activities
UNK	Unknown NACE activity

Source: EUROSTAT High-tech industries and knowledge-intensive services data.

The levels chosen to develop interpretative models of the changes in the structure of employment were as follows:

A-B	Agriculture, Forestry and Fishing, Mining and Quarrying;
D-F	Primary energy sources;
C_HTC	High-technology manufacturing;
C_HTC_M	Medium high-technology manufacturing;
C_LTC	Low-technology manufacturing;
C_LTC_M	Medium low-technology manufacturing;
KIS	Knowledge-intensive services;
LKIS	Less knowledge-intensive services.

Where:

$$C\_HTC + C\_HTC\_M + C\_LTC + C\_LTC\_M = CKIS + LKIS = [G - U]$$

And defining with:

$$\begin{aligned} AGGR1 &= C\_HTC + C\_HTC\_M + KISAGGR2 \\ &= [A - B] + [D - F] + C\_LTC + C\_LTC\_M + LKIS \end{aligned}$$

the two aggregates of high (*AGGR1*) and low (*AGGR2*) technology intensive sectors.

Then:

$$[A - B] + C + [D - F] + [G - U] = \text{Total employment for the selected areas}$$

Or

$$AGGR1 + AGGR2 = \text{Total employment for the selected areas}$$

At territorial level, the analysis was conducted with reference to the two North-Centre and South districts of Italy in the period 1994–2018.

3 Walking Through Sample Data

Figure 1 shows the trend of employment in the aggregate of knowledge-intensive service (KIS) sectors recorded in the observation interval and for the two macro-areas of North-Centre and South Italy.

Over the course of the 24 years, employment in KIS sectors has grown in both areas, but growth in the North-Centre appears significantly faster than in the South. The gap between the two areas widens especially in the period following the 2008 crisis.

In particular, after the peak recorded in 2008, employment in the South ceased to grow, remaining stable at around 2.2 million people (1.3 million, starting level in 1994). On the contrary, in the North-Centre, employment continued to grow, although with a clear reduction in speed (5.8 million people in 2018 and 3.4 in 1994).

The analysis of the composition of the work, expressed in terms of the shares of the KIS sectors in the respective area totals, whose historical trends are shown in Fig. 2, highlights some significant elements of the dynamics of the distribution of work in the two areas:

- Until 2008, the shares of the two areas were substantially similar both in value (from 24% in 1994 to about 30% in 2007) and in dynamics;

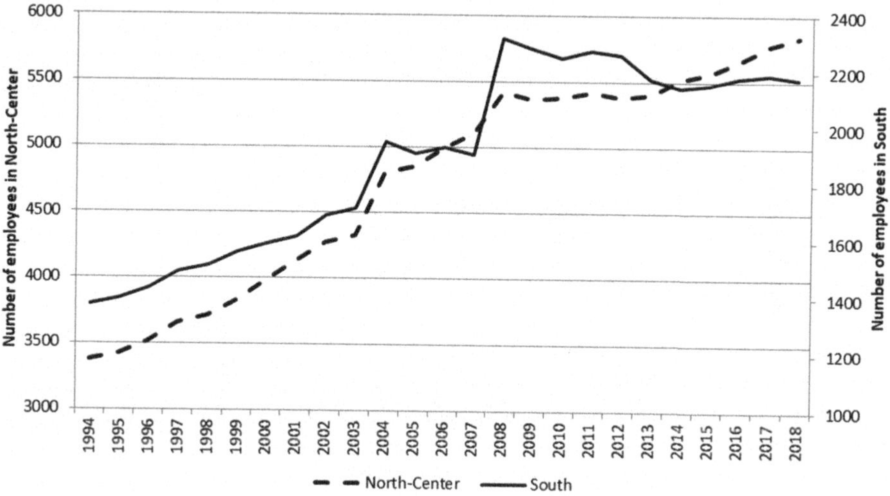

Fig. 1 North-Center and South trends in KIS sector (number of employees). Source: Our treatments on EUROSTAT High-tech industries and knowledge-intensive services data

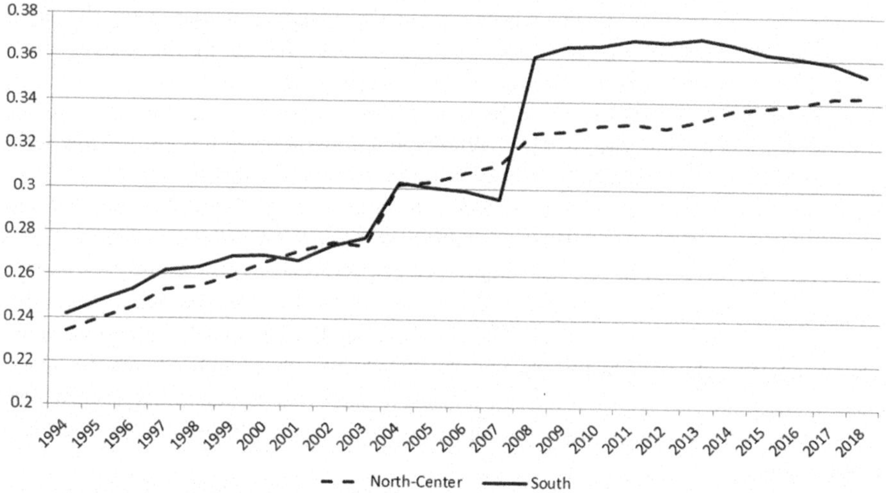

Fig. 2 North-Center and South trends in KIS sector (shares of total employment). Source: Our treatments on EUROSTAT High-tech industries and knowledge-intensive services data

- In 2008, there was a sharp rise in the sector's share of the South (from around 29% in 2007 to around 36% in 2008), as opposed to stable growth in the North-Centre;
- Between 2008 and 2018, the share of the KIS sector in the North-Centre continues to grow with the same trend, while the share in the South stabilizes at around 35% over the 10 years.

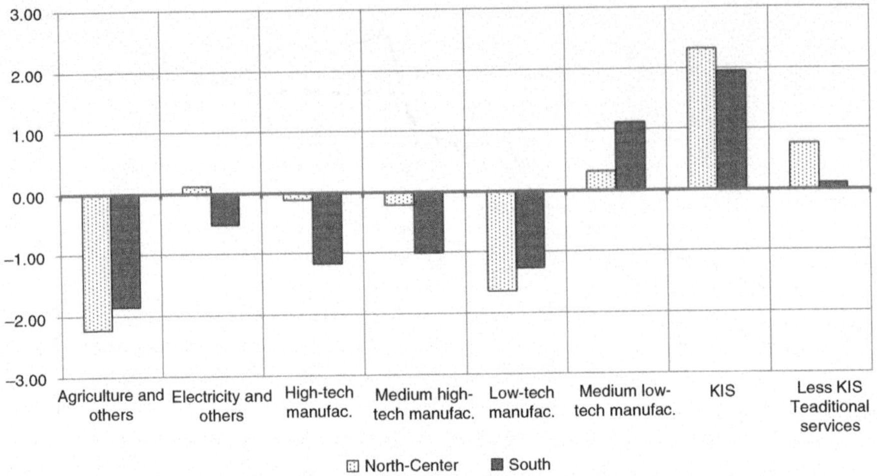

Fig. 3 Employment annual average change rates 2018–1994 (%). Source: Our treatments on EUROSTAT high-tech industries and knowledge-intensive services data

Widening the analysis to the whole economic system, articulated for the eight macro-sectors listed in the previous paragraph, the change in the employment sectorial composition appears even more evident (Fig. 3).

In absolute value, employees in both areas are reduced or remain stable in the primary sectors (Agriculture, Energy, Water), in Construction, in high and medium-high technology manufacturing, in low technology manufacturing.

Employment growth can be identified only in knowledge intensive services (Air transport, Finance, Insurance, Business activities, Computer, Research, Health), less technology intensive services (all other services), and in medium-low technology intensive manufacturing (coke, petroleum, plastic, non-metallic and metal products).

The significant difference in average growth rates between KIS, on the one hand, and Less KIS and medium-low technology intensive manufacturing, on the other, leads to the conclusion that the labor sharing model that is becoming established, sees a consistent and growing specialization in the technology-intensive services sector alone.

This conclusion is confirmed by analyzing the percentage change in employment shares over the 24-year interval (Fig. 4).

The only production sector that sees its share of the economy in both areas growing significantly is the high-tech services sector (KIS). Moreover, this evolutionary trend, with a leading role acquired by the advanced services sector, is common to several other European countries, such as Spain, France, Germany, UK (Lo Cascio and Bagarani 2019).

Summing up, the path of development of the labor market is characterized by a process of substitution, similar to that of other competing European countries, in

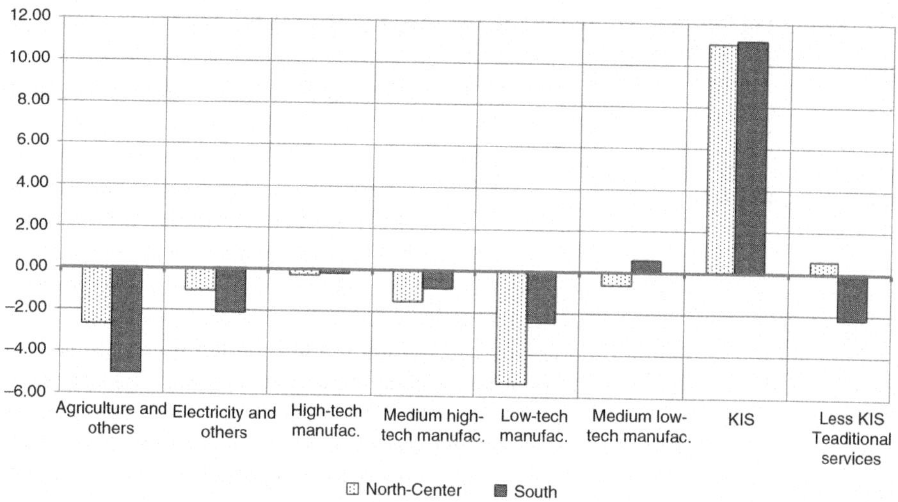

Fig. 4 Δ% of the employment shares between 1994 and 2018. Source: Our treatments on EUROSTAT high-tech industries and knowledge-intensive services data

which, for Italian sub-areas, the employment in technologically advanced services tends to prevail over all other forms of work. On accounting ground, high technology employment in manufacturing loses his weight, on total employment, because the outsourcing of the productivity chain, coupled with robotization and mechanization processes, balancing the more continuous growth of Knowledge Intensive Services.

Finally, even if the structure of employment in Italy shows a similar behavior, that happens, however, at a slower pace of convergence to the new paradigm, at least with respect to its main competitors (Germany, France, Great Britain and, in perspective, Spain).

At first glance, the South economy appears to be considerably compromised, with a significant loss of weight on the total national economy, with a few exceptions mainly related to the resilience of the KIS sector and the ongoing transformations in agro-food production.

In the seventies, several studies for technologies competing penetration, for instance, of railroads *versus* horse transport, highways *versus* railroads, airlines *versus* highways, high-speed trains *versus* airlines, were carried out by Marchetti (1988). The use of logistic functions in these studies proved to be useful even for forecasting further competition and takeover by the new competitors.

We think that this reference is the simplest way to catch the model presented in subsequent paragraph.

4 The Models by Sector and by Region in Brief

The relative position in time of the areas considered may be captured by a naïve model in analogy with the typical techniques used since the seventies for the study of substitution processes and market penetration of the different energy sources (Marchetti and Nakicenovic 1979; Marchetti 1980). "One general finding is that almost all binary substitution processes, expressed in fractional terms, follow characteristic S-shaped curves, which have been used for forecasting further competition between the two alternative technologies or products, and also the final takeover by the new competitor." (Marchetti and Nakicenovic 1979, page 1).

In our case, the S-shaped curves are used to analyze the competition between two labor market alternatives and to identify which alternative can possibly take the lead in the labor market for the next few years. The estimation of the substitution processes took place through logistic functions, regressing logarithm odds ratios over time, for aggregate, sub-groups up to elementary groups of two sectors. A procedure for going back to the aggregation of elementary and sub-groups allow us obtaining estimates that satisfy the constraint: sum of shares equal 1. The selection among different partition paths is carried out utilizing the MLE system R-squared. The following Diagram 1 shows the two-way path of nested elementary and aggregate groups.

Where: *s* indicates the individual sectors, *a* and *b* indicate the groups and *TE* indicates the employment for the total of the sectors in the country.

Summing up, the logistic functions of the eight groups of sectors have been estimated in two macro-areas, North-Center and South of Italy, by minimizing a loss of information function of the odds ratios of employees in a sector as a function of

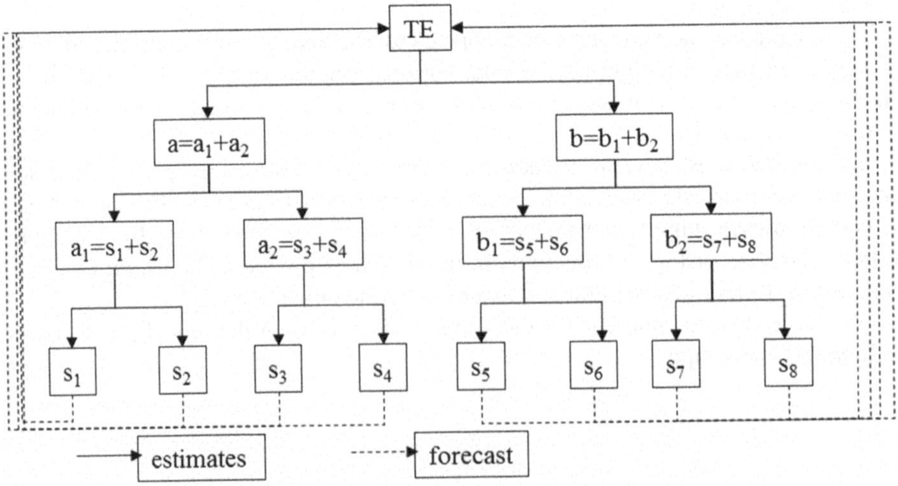

Diagram 1 Path of substitution processes

time (**model 1**), following grouping procedure described in Lo Cascio and Bagarani (2019).

$$\ln\left(\frac{q_i}{1-q_i}\right) = \alpha_i + \beta_i * t + u_{i,t} \tag{1}$$

then

$$\widehat{q}_i = \frac{1}{1 + e^{-\widehat{\alpha}} * e^{-\widehat{\beta_i} * t}} \tag{2}$$

since

$$exp\left[ln\left(\frac{q_i}{1-q_i}\right) + ln\left(\frac{1-q_i}{q_i}\right)\right] = 1 \tag{3}$$

then

$$\sum_i q_i = \sum_i \widehat{q}_i = 1 \tag{4}$$

For k nested grouping of $i \subseteq \Sigma_k i$.

The same model was applied on the odds ratios of employees in each macro-area as a function of time estimating Eq. (1) for each sector, including the total (**model 2**).

5 Results

The analysis of the sample data led to identify the existence of a common labor market penetration behavior between the two Italian macro-areas.

This common behavior is based on the affirmation of technologically qualified employment in the service sector, replacing employment in all other sectors.

The interpretation of the common behavior through the above described models (model 1 and model 2) based on the estimation of logistic functions on the eight production sectors confirms in both macro-areas:

1. the existence of this evolutionary process;
2. the fact that this behavior is characteristic for both areas; and
3. that this behavior is the same as those adopted by the most important European countries in the same time interval.

A first application of Model 1 was made on employment observations in the two high and low technology intensive sectorial aggregates (AGGR1 and AGGR2).

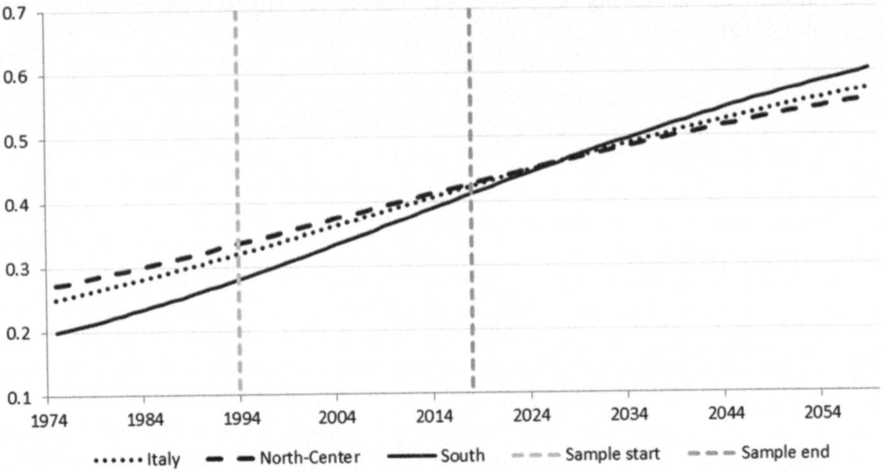

Fig. 5 Model 1—Logistic shares for the highest knowledge-intensive sectors (AGGR1). Source: Our treatments on EUROSTAT High-tech industries and KIS data

Two points emerge from the analysis of the results of Model 1, obtained on the highest technology intensive aggregate (AGGR1)[1]:

(a) although starting from estimated lower shares, the South Italy shows a dynamics of employment substitution greater than the North-Centre (Fig. 5);
(b) for both macro-areas, the inflection points of the logistics are within the sample range (Fig. 6).

In the North-Centre Italy model, the loss of momentum occurred in 2007, while in the South Italy model in 2012.

The trend analysis of the prime derivatives not only confirms the higher speed of the Southern Italy model, but also shows the convergent trend of the two models over the long term.

In other words, Southern Italy is experiencing faster phases of exponential growth (before the point of inflection) and saturation (after the point of inflection) than Northern-Central Italy.

Going down to the analysis of the results of Model 1 for each of the eight macro sectors, the following aspects are highlighted (Figs. 7 and 8):

1. Market shares of Agriculture, Forestry and Fishing, high-technology, medium high-technology and low technology Manufacturing tend to decline over the long run in both the macro-areas;

[1]Since the sum of the two aggregates is equal to the total number of people employed in the two areas, the representation of the aggregate with the highest technological intensity was chosen, the other one being specular.

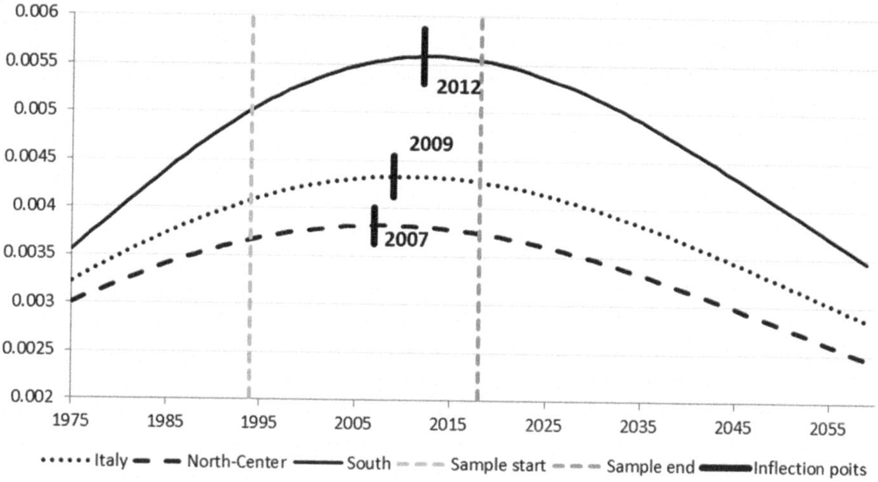

Fig. 6 Model 1—Point derivatives of logistic functions (AGGR1). Source: Our treatments on EUROSTAT High-tech industries and KIS data

2. Market shares of knowledge intensive Services grow considerably over the long run in both the macro-areas;
3. Market shares of Energy, Water and Constructions and less knowledge-intensive Services tend to decline over the long run in the South Italy but tend to grow weakly in the North-Centre Italy;
4. Market shares of medium-low technology Manufacturing tend to decline over the long run in the North-Centre Italy but tend to grow significantly in South Italy.

The greater substitution dynamics emerging for Southern Italy, in the case of the KIS sector, is probably due to the combined effect of two factors:

1. A higher employment dynamic in the KIS sector;
2. A higher rate of expulsion of labor force from the sectors with lower competitiveness (what we could call "zombie" sectors).

Model 2 analyses the processes of substitution between areas, within the same sector, and no longer between sectors, within the same area, as was the case in Model 1 (Figs. 9 and 10).

The employment shares of the South, compared to the other macro-area, grow for Agriculture, Forestry and Fishing, low-technology Manufacturing (Agro-food, Textile, Wood chains and some other sectors), medium low-technology Manufacturing and KIS sectors.

On the opposite, the employment shares of the South decrease for all other sectors.

Within the sample and out of the sample, trajectories of the model's 1 and 2 shares define a framework of structural evolution in time of labor—by knowledge content, activity, and Italy and Nord-South's partitions—consistent with current data, and their above qualitative appraisal.

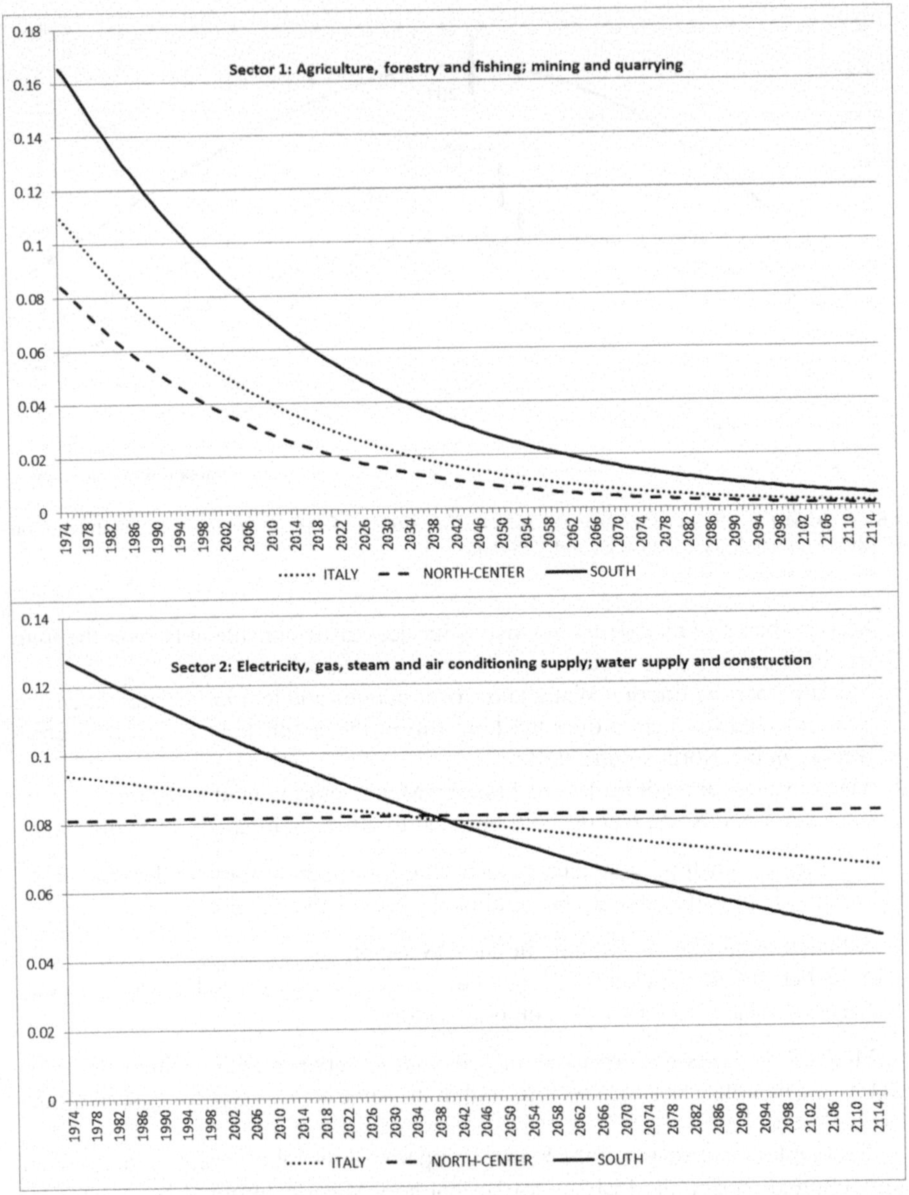

Fig. 7 Model 1—Logistic functions (1). Source: Our treatments on EUROSTAT High-tech industries and KIS data

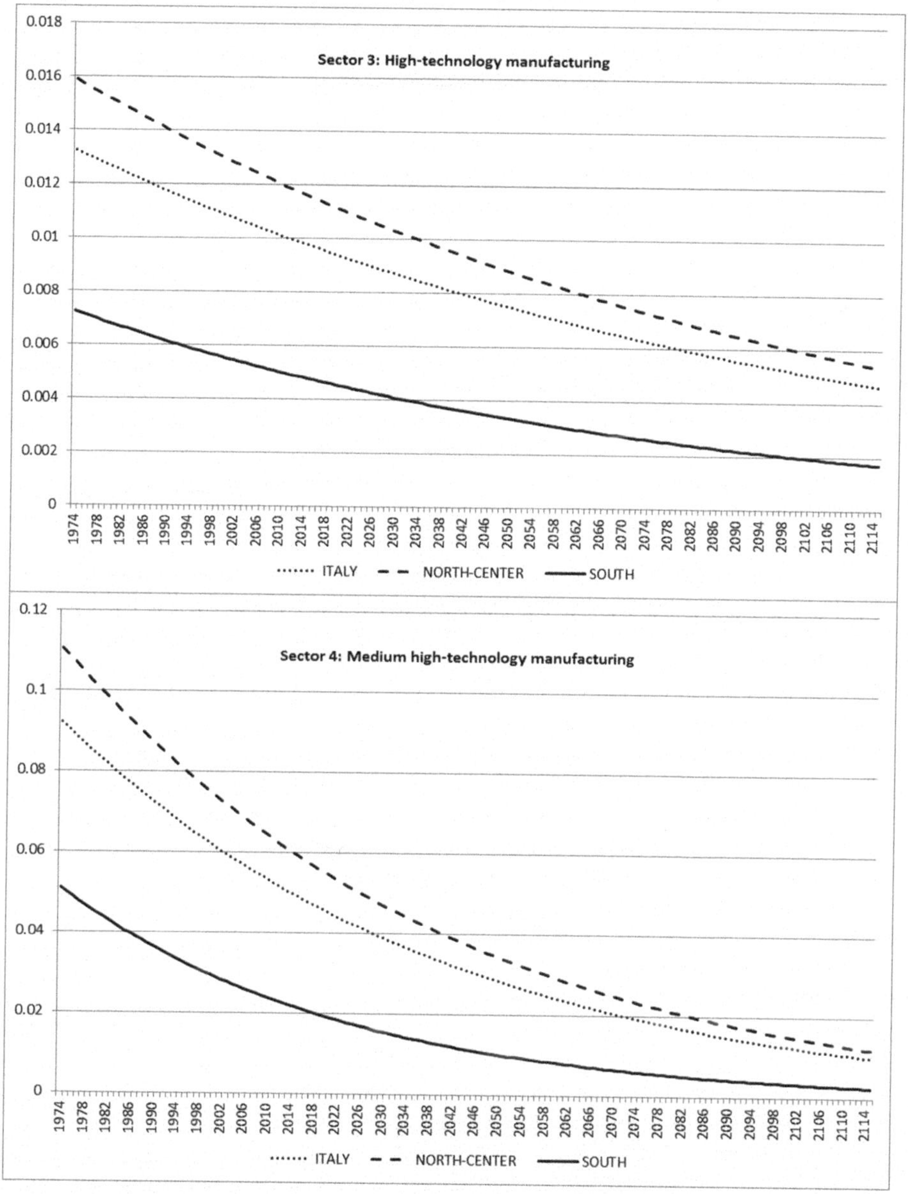

Fig. 7 (continued)

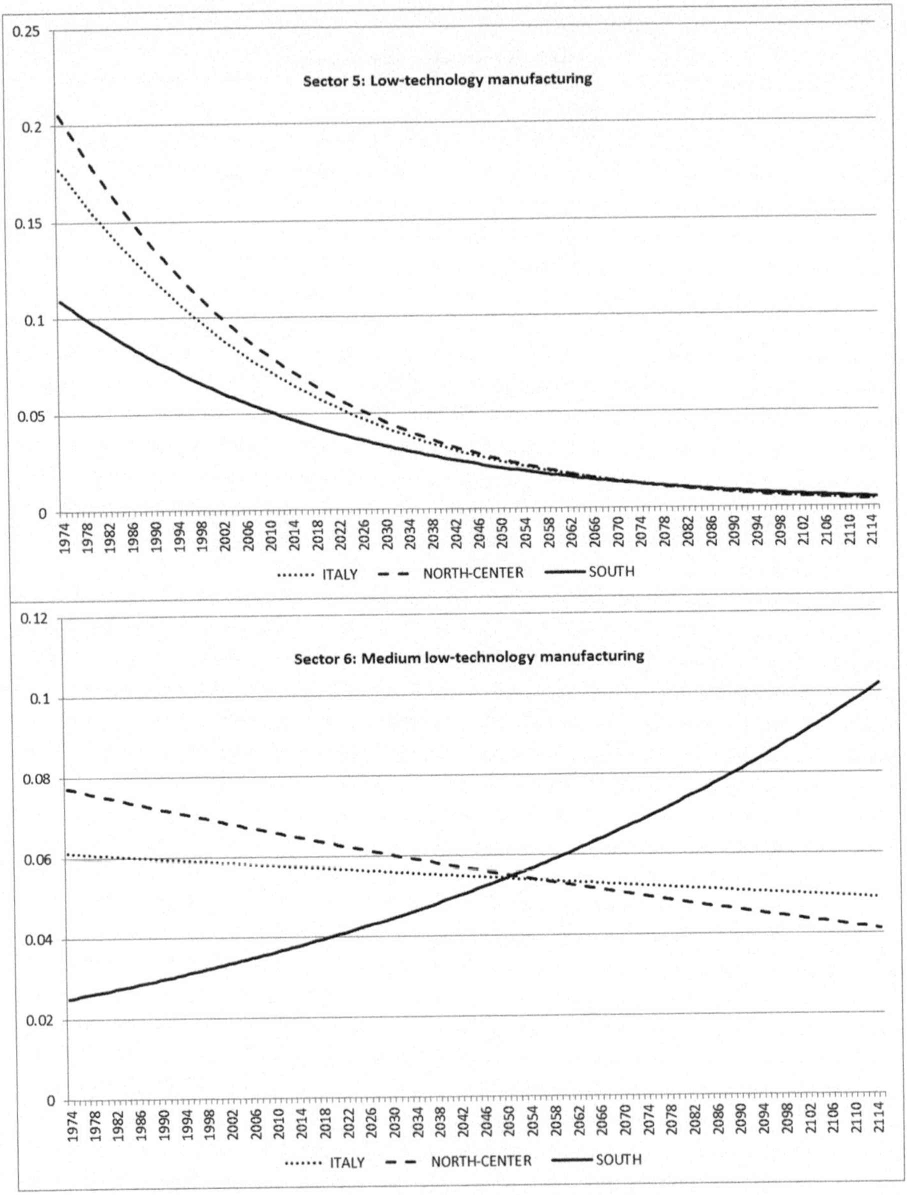

Fig. 8 Model 1—Logistic functions (2). Source: Our treatments on EUROSTAT High-tech industries and KIS data

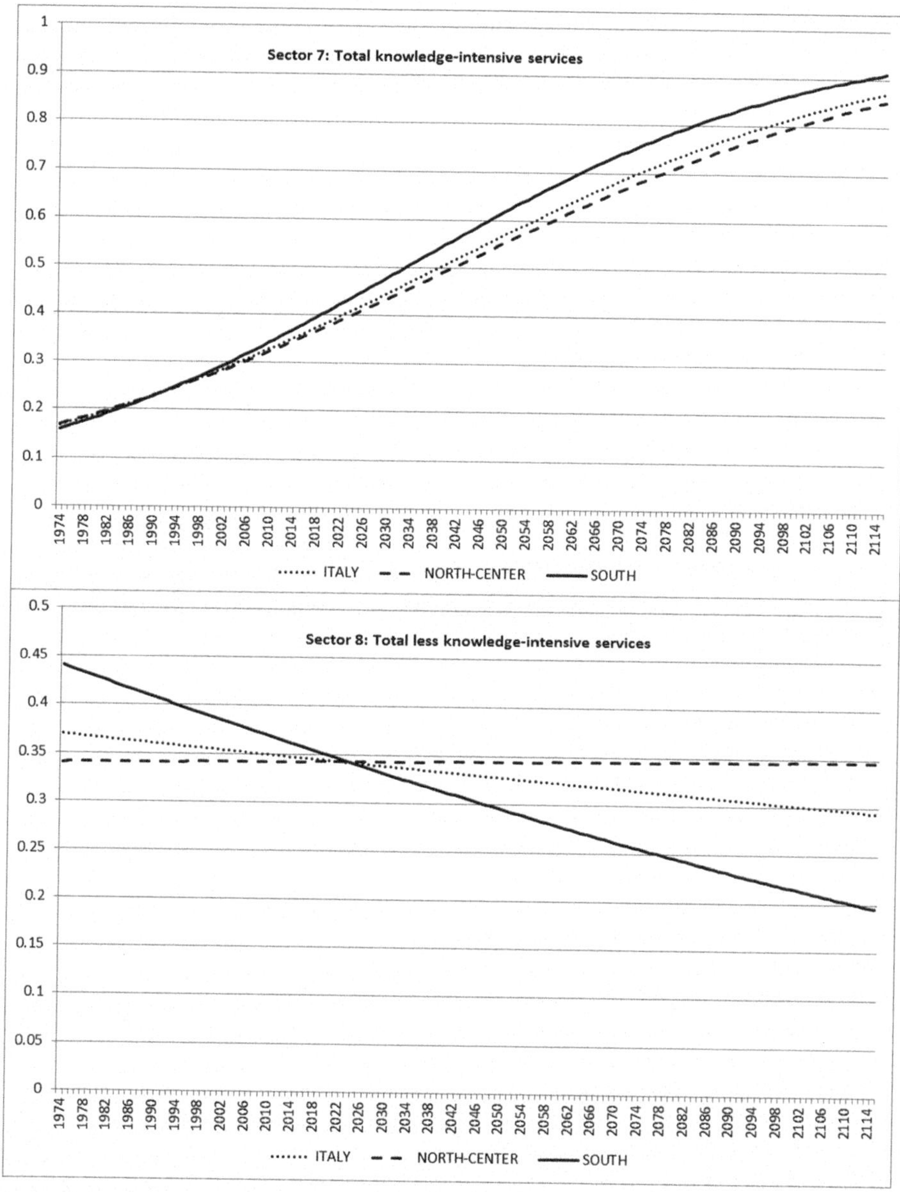

Fig. 8 (continued)

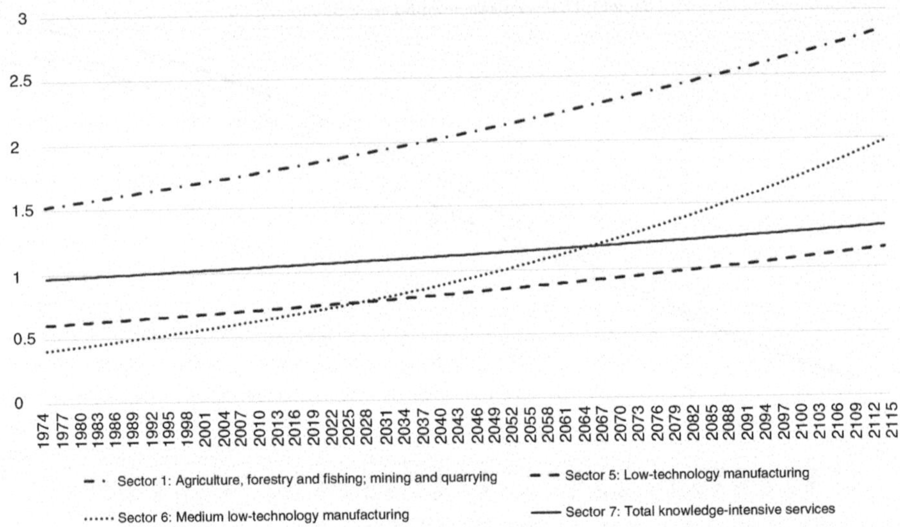

Fig. 9 Model 2—Increasing sectorial shares relative to Total South shares. Source: Our treatments on EUROSTAT High-tech industries and KIS data

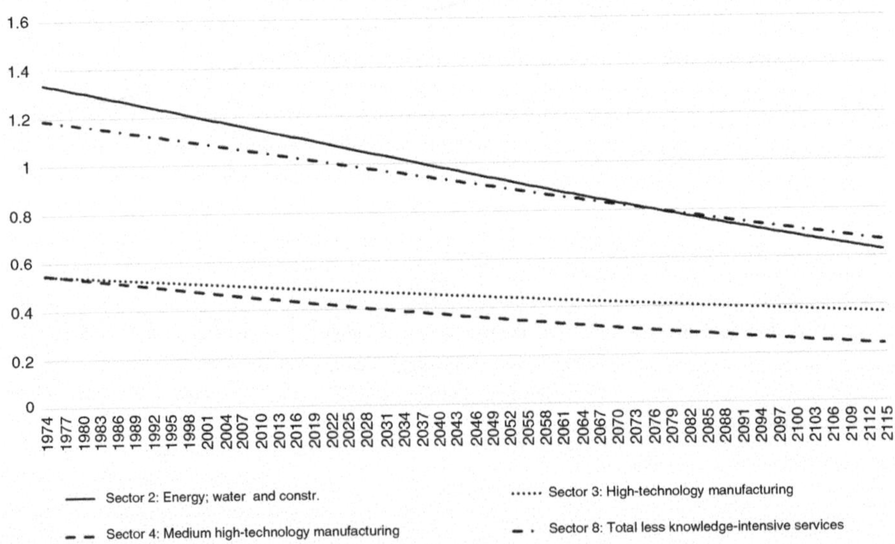

Fig. 10 Model 2—Decreasing sectorial shares relative to Total South shares. Source: Our treatments on EUROSTAT High-tech industries and KIS data

6 Remarks and Conclusions

Summing up the results from model 1 and 2, the following points emerge:

- A structural model similarity within the Italian macro-areas: KIS sectors show an increasing share, in both the two national macro-areas (North-center and South).
- At least in the medium time, we can expect a decline in overall economy and employment in South of Italy.
- At the same time, the momentum of structural adjustment in the South is higher than the North-Center and the logistic of knowledge intensive services (KIS) is expected to be more similar to the European countries than the other model (North-Center).
- Manufacturing sectors with low and medium-low technology-content show a sort of recovery in competiveness for Southern Italy. To same extent similar behavior can be captured by agriculture, whose productivity growth bridges the gap with the Centre-North.
- Mainly, the share of "zombie" activities (low tech services and manufacturing) declines faster, the base manufacturing productions (Steel and Oil refining), the great plants heritage of the past, being an exception.
- The panorama of the South is that of a depleted area but, at the same time, with some favorable perspective of structural adjustment, at least in the medium long term.
- The leopard spot landscape of the southern economy—which emerges from recent studies at the micro level—like that of De Vincenti (2020), and SVIMEZ (2019)—confirms the results of our study, being the last one a sort of framework in which to put the micro-level evidences.

References

Brian Arthur, W. (2013). *Complexity economics*. Oxford: Oxford University Press.

Brian Arthur, W. (1989). Competing technologies, increasing returns, and lock-in by historical events. *The Economic Journal, 99*, 116–131.

De Vincenti, C. (2020). *Cosa frena e cosa spinge la crescita del Sud*. Lavoce.info.

Jones, C. I., & Romer, P. M. (2010). The new Kaldor facts: Ideas, institutions, population, and human capital. *American Economic Journal: Macroeconomics, 2*(1), 224–245.

Kaldor, N. (1961). Capital accumulation and economic growth. In F. A. Lutz & D. C. Hague (Eds.), *The theory of capital* (pp. 177–222). New York: St. Martins Press.

Lo Cascio, M. (2019). *L'Europa e l'economia del cortile*. Roma: Aracne.

Lo Cascio, M., & Bagarani, M. (2012). Economic space trajectory thorough different regional growth models. In M. Bagarani (Ed.), *IL governo delle Regioni e lo sviluppo economico* (pp. 93–110). Edizioni dell'Orso.

Lo Cascio, M., & Bagarani, M. (2018). Incoming labor-product Society and EU Regional Policy. In L. Paganetto (Ed.), *Getting globalization right. Sustainability and inclusive growth in the post Brexit age* (pp. 241–263). Cham: Springer.

Lo Cascio, M., & Bagarani, M. (2019). Great European crisis: Shift or turning point in job creation from job destruction. In L. Paganetto (Ed.), *Yearning for inclusive growth and development, good jobs and sustainability* (pp. 189–202). Springer.

Marchetti, C. (1980). Society as a learning system: Discovery, invention, and innovation cycles revisited. *Technological Forecasting and Social Change, 18*, 267–282.

Marchetti, C. (1988). *On society and nuclear energy. Part I and Part II*. IIASA.

Marchetti, C. (2003). Logos, il creatore di imperi. *Systema Naturae, 5*, 307–321.

Marchetti, C., & Nakicenovic, N. (1979) *The dynamics of energy systems and the logistic substitution model*. IIASA Research Report December 1979, Laxenburg.

Romer, P. (1986). Increasing returns and long-run growth. *Journal of Political Economy, 94*(5), 1002–1037.

SVIMEZ. (2019). *L'economia e la società del Mezzogiorno*. SVIMEZ.

Artificial Intelligence, Its Corporate Use and How It Will Affect the Future of Work

Jacques Bughin

Abstract In the current debate over the Future of Work, there is little discussion about how firms anticipate the evolution of their demand for labor and the related mix of skills as they adopt Artificial Intelligence (AI) tools. This article contributes to this debate by leveraging a global survey of 3000 firms in 10 countries, covering the main sectors of the economy. Descriptive statistics from the survey are complemented by econometric analyses of corporate labor demand decisions. The findings are four-fold. First, those are still early days in the absorption of AI technologies, with less than 10% of companies investing in a majority of AI technologies and for multiple purposes. Second, if an aggregate portion of firms anticipates reducing employment as a result of adopting AI technologies, as many other companies anticipate labor growth or reorganizing employment. Third, this reallocation picture holds true when we examine further demand by labor functions and skills, with talent shifting toward more analytic, creative, and interaction skills, and away from administrative and routine-based functions, in line with past trends of skill- and routine-biased technological change. Fourth, a novel to the literature on Future of Work, econometric results on employment change highlight that employment dynamics are driven by related spillover effects to product markets. Higher competition, larger expectations of market (share) deployment may counterbalance negative automation effect on employment dynamics.

This article is based on a series of work launched at the McKinsey Global Institute where the author was a director until end of Dec 2019. This article is a personal contribution and does not engage any of the institutions mentioned. All errors and omissions are the author's own.

J. Bughin (✉)
Free University of Brussels, Brussels, Belgium

Solvay Management School, Brussels, Belgium

Marquard Media, Zug, Switzerland

Portulans Institute, Washington, DC, USA

Retired from McKinsey & Company and McKinsey Global Institute, Brussels, Belgium

L. Paganetto (ed.), *Capitalism, Global Change and Sustainable Development*, Springer Proceedings in Business and Economics,
https://doi.org/10.1007/978-3-030-46143-0_14

Keywords Artificial intelligence · Derived labor demand · Product market competition

1 Introduction

Recent advances in the field of artificial intelligence (henceforth, AI) have led to public fears that these technologies will substitute a large part of job occupations (Brynjolfsson and McAfee 2014, Brynjolfsson et al. (2017) or Nubler 2016). This fear is being fueled by companies announcing their intention to replace groups of workers by smart algorithms and/or robots.

At the same time, a recent stream of academic work has strengthened the vision of a "workless future." In their seminal work, Frey and Osborne (2013) calculated that 47% of all US jobs are at risk of being automated by the rise of AI technologies. Another recent study by Acemoglu and Resterpo (2017) claims that every robot makes as many as eight jobs obsolete. Follow-up research by same authors (Acemoglu and Restrepo 2018) suggests that new technologies may fuel more automation than new jobs, creating inequality risks between workers, and likely pressure on total jobs. Susskind (2020), leveraging Zeira (2018) develops a theoretical model of smart capital augmentation which is fully substitutable to jobs at high-wages, leading to an extreme scenario in which "wages can only decline to zero" to secure workers employability. Empirically, a recent study linking labor productivity and employment in a sample of large OECD countries by Autor and Salomons (2017) also suggests that higher productivity at sectorial level (often driven by technology innovations) is associated with *decline* in employment in the same sector.

Such fears are not new. Already at the time of the first Industrial Revolution, renowned thinkers such as John Stuart Mill and David Ricardo conceded the possibility of unemployment. Given the rise of manufacturing and its need for workers, however, the concern shifted quickly to issues around wages, which stagnated for 50 years until the middle of the nineteenth century—a "pause" noted by Engels.

The Great Depression brought a revival of concerns. John Maynard Keynes (1931) wrote his famous essay *Economic Possibilities of our Grandchildren*, predicting that by 2030, the "most pressing problem in developed economies would be how to fill our leisure time." Today, the top ten US firms by market cap employ 30% fewer people than the top ten firms in the 1960s.[1] Featuring among those top ten are the so-called GAFAs—Google, Amazon, Facebook, and Apple—which are able to generate every $1 of value with five times fewer employees than the largest US firms 50 years ago.

Some research brings however more nuance to a scenario of large unemployment. Gregory et al. (2016) shows that tasks rather than jobs are being replaced, and technology also created new demand and jobs. Atkinson (2013) concludes that, at

[1]Discussion with Mark Purdy from Accenture research at the G-20Y in Evian, Sept.

Table 1 Adoption stage of AI technologies, 2017, % of large size firms worldwide

	AI technologies:					
Adoption status	Machine learning (%)	Computer vision (%)	Language processing (%)	Robotics (%)	Virtual agents (%)	Robotics process automation (%)
Not aware	29	24	5	12	42	90
Not yet invested	29	39	43	12	29	2
Piloting	15	14	18	14	11	2
Adopted one use case	13	12	18	28	10	2
Adopted many use cases	13	11	14	33	10	2

least looking backwards, there is no single decade in the United States, from 1850 through 2010, in which the adoption of technology did not destroy employment *more* than it was responsible for creating new jobs. A McKinsey Global Institute (2017a)'s research across 46 countries has attempted to match *technical* capabilities of AI with those of humans —for example a virtual assistant system answering questions versus the task being done by a call-center agent. Recognizing that typically jobs are composed by many tasks, it is found that automation technology would likely more affect the mix of activities *within a job* than it will threaten to replace an entire occupation. On average, in developed countries, the study finds that 25%–30% of existing jobs runs the risk of 70% of their tasks being automated. Further, the shift in tasks will be felt in more routine-based ones than in tasks requiring social and creative skills.[2] Those findings have been corroborated by parallel research at OECD (2017).

Likewise, absorption of technology takes time. Looking at a wide set of technologies in the last century, the median time of complete diffusion has been close to 40 years, even if diffusion speed has accelerated lately for digital technologies (Comin and Hobijn 2010). Our survey of large companies (see Table 1) confirms that AI adoption is in its early day. There are also large bottlenecks in digital assets and skills needed within the majority of incumbent firms that may lead to a somewhat slow diffusion of AI technologies (Bughin and van Zeebroeck 2018).

This paper contributes to the debate on a "jobless future" through the *demand side of the labor market,* as currently corporations account for about 90% of the jobs in developed countries. Through this lens, we hope to discover micro-findings on the dynamics of employment in relation to the corporate diffusion of AI technologies that are both consistent and new.

In effect, we bring together multiple streams of literature, on top of labor economics literature. Regarding the IT/IS literature, it is already clear that

[2]See *A future that works: Automation, employment, and productivity*, McKinsey Global Institute, January 2017.

employment will follow gains of productivity arising from absorption of technologies (see Brynjolfsson and Hitt 2003), while technology adoption dynamics should mostly affect the mix of skills rather than total level of employment (Machin 2001). One reason for this is also that corporate returns to technology investment are often only attractive to the extent that companies invest in the complementary human skills that make technologies operate effectively at scale. As a case in point, Bughin (2016) shows that returns to big data investments are higher than the corporate cost of capital for firms which have also invested in the right pool of big data analysts. In fact, relying on work by the McKinsey Global Institute (2017a, b, 2018a, b), Pissarides and Bughin (2018) conjecture that the most important challenges will be the skill uplift associated with the diffusion of smart technologies, and frictions around this transition is likely one key driver of how labor markets behavior in years to come. Second, regarding the Strategic Management and Industrial Organization literatures, companies invest in technology for a variety of reasons, including greater labor efficiency but also because they wish to *create better or* new products/business models and expand their sales, in relation to product market competition and conduct. Hence, labor demand may increase (pending on how strong and fast competition reacts), as documented during the early diffusion of PC-based technologies (Spiezia and Vivarelli 2000), or in subsequent research by Garcia et al. (2002) and Peters (2004).[3]

This article looks at three hypotheses. The first one relates to the previously-noted fear that *AI will significantly reduce number of jobs*. This fear may be overblown, but has some truth as well. As we will see hereafter, our survey results suggest that, in aggregate, more than half of firms anticipate changes in employment *but if a sizable share of companies is considering reductions in employment—it may be only for some categories and a reallocation of skills*.

The second hypothesis is the skills- and routine-biased technological change hypothesis of a shift to more complex skills and less *routine-based jobs* [see Autor (2015) or Goos et al. (2015)]. If one takes the first wave of IT technology deployment—that is, mainframe and PC—Handel (2016) suggested that IT accelerated the ongoing shift towards jobs requiring higher education by roughly 50%. When General Motors started to adopt the first generation of automation technologies in the car manufacturing industry in the 1980s, problem-solving capabilities for skilled workers rose by 40% (and fell by 10% for unskilled, blue collar workers), as did tasks with higher memory, accuracy and concentration skills (Milkman and Pullman 1991). As detailed hereafter, *we find support for the skill-bias hypothesis in our survey. In particular, we find evidence for employment increases in big data*

[3] As a case in point, let us consider the Associated Press news agency, which used to deliver reports on large corporations using 65 journalists in its newsroom. With AI technologies, the company quickly managed to automate the production of simple stories of quarterly earnings for 10 times as many small companies in the long-tail. This output gain was not done at the expenses of reporters; the in-house reporters did not lose their jobs, but were instead redirected to write longer research article on business trends as a major latent demand spotted by the company. See for detailed reference, the article by Ramaswamy, S (2017) at https://hbr.org/2017/04/how-companies-are-already-using-ai.

analytics and for occupations requiring interpersonal skills for corporations already well vested in AI. Increase in AI investment is also positively correlated to better odds ratios in categories such as leadership and creative design.

The third hypothesis is that the effect of AI automation on employment will ultimately depend on product market overspill. These include new products and its interaction with a company's profitability, in line with the oligopoly theory of firm-induced labor demand. Our econometric *results are the first, to the best of our knowledge, to demonstrate that the effect of AI technology on labor demand must take into account those spill-over effects on product markets.*

This article is structured as follows: Sect. 2 describes our survey, as well as our definition of AI and the hypotheses we test. Section 3 presents the descriptive results, while Sect. 4 is concerned with the econometric analysis of labor demand by type of firms and firm pace of adoption of AI technologies. Section 5 concludes.

2 Background

2.1 AI Definition

Our definition of AI in the article (and in the survey questionnaire) is one of technologies being technically able to *mimic cognitive human functions.*[4]

Consider Amazon's Kiva robots. Today, they are already able to handle parcels faster than humans and in a space that is half the size of traditional human–heavy logistics centers, leading to significant efficiency gains, and with return of investments easily a multiple of cost of capital (see McKinsey Global Institute 2018b). Likewise, virtual assistants such as IPSoft's Amelia are able to handle customer care much faster, and with more consistent positive responses, than humans, reducing cost of care by more than 50%, and improving quality of interactions with users. The ability of AI to perform and self-learn obviously depends on big data, and on the use of powerful algorithms such as deep learning. Alphabet's DeepMind reported that it improved the overall power usage efficiency of Google's data centers by 15% after placing an AI program similar to a program taught to play Atari games in charge of managing a data center's control system.[5]

Based on actual evidence of technologically readiness, we have surveyed the use of five types of technologies (computer vision, language processing, robotics, robotic process automation, and virtual agents) in the market place. We add to the list deep-learning techniques (and their derivatives), as most of the above technologies usually rely on deep machine-learning algorithms to deliver their results.

[4]Substitution may arise when, furthermore, the economics are attractive to replace human capital for example.

[5]https://www.infoq.com/news/2016/07/deepmind-cooling-pue

2.2 *Survey Collection and Highlights*

2.2.1 Sample

We leverage a C-suite executive survey conducted in the spring of 2017, covering 10 countries and 14 sectors. The survey was commissioned externally to a major research firm, covering topics such as awareness and use of a set of AI technologies, returns to AI investment, as well as impact on strategy and labor resources allocation by skills and company functions. In total, there were about 25 questions to answer, for an average time to fill of less than 20 min, in order to maximize take-up rates and adequate responses. The survey was administered online. The survey questionnaire is accessible in the appendix of other research report by Bughin et al. (2017).

We received 3073 fully completed and validated set of responses out of a original sample of 20,000 firms stratified to reflect both firm size distribution (small, medium, large) as well as sectorial contribution in added value to each country's GDP. The answer rate is a relatively good answer rate (more than 15%) from a total random sample of companies.

The ten countries we focused on were the United States, Canada, the five largest European countries and Sweden, China, Japan and South Korea. We picked those countries as they are the largest contributors to world GDP, are the most digitally-advanced, and have recently scaled their investments in AI recently.[6] The largest portion of answers came from the United Kingdom (12%) followed by the United States. The country with the fewest answers was Sweden (5%). 27% of firms were very small firms, i.e. with fewer than 10 employees, while 7% of the sample includes firms with more than 10,000 employees. The sample covers service, agriculture, and industrial sectors, from professional services (14% of our final sample) and high tech (10%) or retail (8%), to travel and tourism (4%), automotive/assembly (4%), or the education sector (5%).

The responses we received were tested for absence of bias by industry—specifically we tested whether they were any difference in our sample of answers with the original target of firms, in terms of mean difference in key financial metrics of respondents and non-respondents (revenue, revenue growth, profit and profit growth). We used simple one-way test per financial metric, as well as a multivariate logit model of having answered or not, linked to all the financial metrics (see Whitehead et al. 1993). We could not find statistical difference in answer rates. Finally, we tested for some self-reported biases. This was originally minimized by randomizing questions order in ten subs-samples; those sub samples were then checked for any, and did not find, bias in responses. We nevertheless checked for systematic responding (either extreme, or only middle answers). We spotted 122 answers, or 4% of answers regarding companies whose difference in answers by category of the questionnaire (AI awareness, AI impact on profit, AI impact on

[6]For statistics, see https://www.thetechedvocate.org/six-countries-leading-the-ai-race/ and https://qz.com/1264673/ai-is-the-new-space-race-heres-what-the-biggest-countries-are-doing/

Table 2 Technology variety absorption, 2017, % of firms adopting AI

Degree of penetration of AI technologies among AI-adopting firms	
Adoption of one technology	48%
Adoption of two technologies	24%
Adoption of three technologies	16%
Adoption of four technologies	7%
Adoption of at least five technologies	4%

employment, and on employment mix) were found to be very low (in the bottom 5% in difference in answers across all categories). However, the econometric results are not sensitive to including or not those responses, so we keep our full sample as basis of our results here-after.

2.2.2 Data Highlights

Our survey confirms that like any other technology (Comin and Hobijn 2010), AI adoption may take time to spread. By 2017, diffusion is still in its early days (see Table 1). Only one out of 8 companies report using AI at scale (for multiple use-cases), and a very large set of companies (90%) are not even aware of RPA.

Further, among firms adopting and piloting AI, the variety of absorption is rather narrow. A large portion (48%) has only adopted one of the six technologies surveyed, and only 4% which have already deployed close to the whole set of technologies at scale (see Table 2).

While not reproduced here for sake of space, the sample shows that US and Chinese firms are more advanced in adoption and breadth of diffusion, in line with other research that most buoyant markets in AI investment are in those two geographies (see Bughin et al. 2017). Likewise, sectors that are more advanced in digitization, such as telecom, high tech, and media are already more advanced in AI adoption than other sectors such as construction. Patterns of adoption by technology type is also industry specific; for instance, RPA is twice more often adopted in automotive and assembly than average, while virtual agent technology is most advanced in B2C services such as consumer high tech and telecom (22% versus 12% on average).

Table 3 further provides a picture of the economic motivation for AI adoption, where all sources of motivations are rescaled to 100%. This includes only firms piloting or adopting AI.

This Table 3 provides two critical insights. The first is that, when planning to invest in AI, companies report multiple rationales (3 rationales out the 5 surveyed). Otherwise stated, companies do not generally invest in AI for only a *single* purpose, and usually it is for a mix of efficiency *as well as to facilitate* top line growth. This link to product market is not explicitly discussed in the labor market literature. Looking further at use cases of companies investing in automation of labor, most of those cases have also an augmentation component- e.g. using in-silico AI simulation

Table 3 Rationale for decision to adopt any AI technology, 2017, % of firms piloting and adopting AI

Industry	Output growth			Efficiency gains			
	Market share (%)	Market size (%)	Total (%)	Capital (%)	Labor (%)	Non-labor (%)	Total (%)
High tech	14	21	35	27	24	14	65
Automotive	16	12	28	25	21	26	72
Construction	8	8	17	30	27	27	83
CPG	26	19	45	19	26	10	50
Retail	20	20	40	24	20	16	60
Media	17	29	46	14	17	22	54
Telecom	23	28	52	20	20	8	48
Travel	10	24	34	17	21	27	66
Transport/logistics	24	7	31	24	30	15	69
Financial services	23	23	46	16	30	9	54
Professional services	21	16	38	21	21	20	62
Education	17	20	37	27	16	20	63
Health	20	15	35	24	32	9	65
Energy	21	12	33	24	24	18	67

to support management decisions on market entry; or using algorithms and virtual assistants in marketing to target more appropriate customer segments.

Second, the current debate on the future of work implicitly relies on the issue of labor automation, leading to the question of how, to what degree, smarter capital can replicate human tasks. However, our data demonstrate that *labor* efficiency is only used as motivation for adoption by firms in *only 24% of* cases. Remarkably too, market expansion and market share gains are almost as frequently quoted as labor efficiency (and at higher rates than labor efficiency in four sectors—retail, media and telecom, travel, and education). This clearly suggests that the narrative of labor substitution is possibly too restrictive. In general, firms seem to expand productivity as result of AI adoption by other means than labor, as well as leverage AI to expand their influence on the product market.

Table 4 Expected employment dynamics, 2017, % of firms piloting or adopting AI

Industry	It will reduce our need for employees with total employment levels down (%)	It will reduce employees in some areas, but overall employment may go up (%)	It will not change our need for employees significantly (%)	It will increase our need for employees (%)
High tech	19	32	32	18
Automotive and assembly	23	34	33	10
Construction	25	24	40	11
Consumer packaged goods	19	33	36	12
Retail	16	26	49	9
Media and entertainment	24	25	43	8
Telecommunication	24	32	25	18
Travel and tourism	17	30	41	12
Transport and logistics	22	26	38	14
Financial services	18	32	41	10
Professional services	13	14	66	7
Education	20	20	54	6
Healthcare systems and services	14	28	53	5
Energy and resources	19	33	41	7
Average	19	27	44	10

3 Employment Dynamics- Finding from Descriptive Statistics

Our survey also explicitly has asked on how AI-related technologies has affected or, will impact employment and employment skill mix in the future.[7]

3.1 AI and the Dynamics of Employment

Regarding expectations, survey responses are summarized in Table 4 for companies adopting or piloting AI. 44% of companies do not see an impact, while the portion of

[7]We are rather keen to understand the expectations of firms as the current level of AI diffusion across all technologies is still relatively low.

Table 5 Expected employment dynamics by skill type, 2017, % of firms piloting or adopting AI

	AI impact on employment:				
By functions	Less	More	By skills	Less	More
Back office	28	30	Basic literacy skills	25	33
Front line employees	30	31	Basic numerical skills	27	34
Operations	30	30	Basic IT skills	26	43
Sales and marketing	22	32	Advanced IT skills	21	51
Data and analytics	23	37	Advanced data skills	23	49
Engineers, IT, design	22	41	Critical thinking	24	43
Finance, HR	25	33	Social skills	21	39
Middle management	22	30	Communication skills	20	39
Senior management	19	29	Creative design	23	41
Average	25	33	Craft/technical skills	21	34
			Engineering skills	22	41
			R&D skills	24	47
			Leadership skills	20	33
Average	25	33		23	41

companies reporting a decline in total employment is about 19%, compared with only 10% that report an increase in demand for labor. Interestingly, 27% of companies say that there will be a labor reduction in some occupations, but with a similar employment *increase* in other occupations. In other words, companies tell us that *there is more job re-allocation than total job shrinkage*, in line with our first hypothesis that the effect of AI on employment may be more balanced than feared.

The same conclusion is also drawn if we compared employment changes of AI adopters versus non-AI adopters- in general, those who do not adopt tend to anticipate that AI may reduce more often employment level than those already adopting.

Finally, the pattern of employment dynamics expectations is not necessarily influenced by some sectors. The pattern remains the same at industry level, with the exception of construction, where the dynamics is more on employment reduction. Noteworthy is that sectors that are more advanced in AI are those in which the larger portion of companies expects to both increase and reallocate their labor demand. About half of telecom and high-tech companies surveyed will be ramping or reallocating labor as a consequence of adopting AI- related technologies.

3.2 *AI and Employment Mix Change*

The picture of employment evolution is even clearer if one disaggregates labor demand by occupations and by skills type (see Table 5).[8] The portion of surveyed

[8]Here we show results in aggregate, but the same picture is also visible by industry.

executives who report a reduction in employment is about 25% for almost every function. The highest frequency of decline is visible in functions with *more routine-based* activities such as operations and back-office, while the lowest frequency of decline lies in senior management roles. Further, there are as many companies reporting an increase in the level of employment by function as those reporting a decline. The net balance is less favorable for more routine-based jobs, while functions in data and analytics and IT/Design are more likely to increase than decrease, as already hypothesized in simulations presented in the McKinsey Global Institute (2018a) research on probable skill shits.

The largest portion of firms reporting a decline is for basic literacy and basic numerical skills. 40% of companies will increase any skill, and the largest portion of companies is willing to raise labor demand once again for advanced data and IT skills. This is clearly consistent with our second hypothesis that AI may lead to a skill-biased technological change in employment mix, as happened in the recent past (Autor and Handel 2013).[9]

4 How Artificial Intelligence Technologies Absorption Use Influences Employment Choice: The Econometric Analysis

The above data suggest that the dynamics on employment may be more complex than the narrative that AI will reduce jobs. We formally test this in this section. In the first subsection, we start by specifying a simple derived labor demand that serves as the backbone of our econometric specification laid out in a second sub-section. Then results are reported and discussed.[10]

4.1 The Short-Term Derived Labor Demand as a Function of AI Automation

We consider a profit-maximizing firm facing imperfect product competition, and whose supply is produced though a constant elasticity of substitution (CES) technology with capital and labor as inputs; and weights are directly linked to the extent of how AI is deployed for human task automation. This mimics the recent literature

[9]The same skill-biased tendency is also noticeable in the econometric analysis conducted by Arntz et al. (2016) linking occupations and tasks to the OECD PIACC skill database.

[10]Note that the sample used will concern only firms which are aware of, but not necessarily adopting, AI technologies. We sub-select those firms, and survey responses on the impact on AI are likely not to be largely noisy for those respondents with limited understanding of AI technologies

of looking at the aggregate distribution of task automation to reflect optimal resources mix [see Acemoglu and Restrepo (2018), or in Martinez (2018)].

4.1.1 Product Market Spillover

One important element of this article is to emphasize the link between labor and product markets and how it may affect how AI diffusion influence labor demand.

We thus consider a firm r which maximizes profit, π, while being in competition with other firms, and supplying a product Q_r, with iso-price elasticity of product demand, κ, ($\kappa < 0$).[11] Given competitive interactions, this firm's equilibrium market share is given by, $MS_r = Q_r/Q$ (where Q is total market supply) and product price over marginal costs, $P/MC = \mu_r$ ($=> 1$) is:

$$\mu_r = \kappa MS_r / (MS_r \kappa + 1 + COMP), \tag{1}$$

COMP is a conjectural variation parameter, and lies between [−1,1], and COMP = −1 means perfect competition ($\mu = 1$), while COMP = 0 means that all firms behave as Cournot.

Based on the above, it is easy to show that the growth in μ,($=\phi$), is a positive function of the growth in MS, ΔMS, as well as a negative function of growth in competitive intensity, COMP, ΔCOMP. We note:

$$\delta \log (\mu) = \phi = \phi(\Delta \mathrm{MS}, \Delta \mathrm{COMP}) \tag{2}$$

ϕ in Eq. (2) will turn to be important as it reflects the extent of how AI automation effect on production efficiency is passed-through into profitability of the firm, see infra.

4.1.2 Derived Labor Demand

Now assume that the firm also decides to invest in AI technologies for multiple purposes, in consistency with Table 3. We denote, AIL > 0 if the goal of leveraging AI is labor automation. Likewise, AIE > 0, if AI leads to any other complementary labor productivity gain and AIM(S) > 0, when it concerns the use of AI for enlarging market (share).

The firm produces its output, Q, via a Leontieff function of labor and capital, which is dependent of each of the n AI technologies used. Here we follow Martinez (2018), by stating that there is a one-to-one capital stock that embodies each AI-based technology AIL_j ($j = 1, \ldots n$). This implies that, if it takes workers one unit of time to complete all tasks of the use case in $[0, t_j]$, a worker-AI machine pair

[11] With no loss of generality, we drop the suffix r hereafter.

would produce $1/(t_j - AIL_j)$ goods in one unit of time, and thus the higher AIL_j, the less the need for hiring workers as result of automation arbitrage.

The productivity of a worker-AI pair for attribute j is given by, (with $\gamma > 1$ implies returns to specialization):

$$z(AIL_j) = \left(t_j / {}_{tj - AILj}\right)^{\gamma} \tag{3}$$

Consider further that, the distribution of AI_j, is represented by a beta distribution, with upper bound (AILH) of diffusion, then the firm supply converges to a CES function:

$$Q = T\left((1-\alpha)K^{(\sigma-1/\sigma)} + (\alpha)(L)^{(\sigma-1/\sigma)}\right)^{(\sigma/\sigma-1)} \tag{4}$$

where K, L are aggregate capital and labor at firm level; $\rho = \sigma/1 - \sigma$ where σ is the typical (constant) elasticity of substitution between capital and labor, while the total factor productivity term T and the weight α are endogenized as:

$$T = z(AILH)^{1-\sigma} \tag{5}$$

$$\alpha = z(AILH)^{\sigma-1} \tag{6}$$

If one notates that $\gamma = \rho + \varepsilon$ ($\varepsilon > 1$).

Equations (4)–(6) illustrate how AIL affects factor intensity and in particular, how automation may put pressure of labor. To see that more simply, take the log of (6), that is $\ln(\alpha) = \varepsilon.\ \ln(1\text{-}AILH)$, as well as the special case $\varepsilon = 1$. Then, we find roughly that the labor share is directly linked to increased automation AILH:

$$\alpha = (1 - AILH) \tag{7}$$

This is in line with the idea in the literature that automation exerts a direct decline in labor share (Autor and Salomons 2018).

4.1.3 Employment Dynamics in Function of AI

As said earlier, the derived labor demand must take into account all channels and namely the indirect impact on full supply through the change in μ. In fact, using (2)–(4)–(5)–(7), the first order condition for profit maximization leads to the following short-term employment elasticity that is[12]:

[12]See also Ugur et al. (2016).

$$\tau = (\partial \log L/\partial \text{AILH}) = -\kappa\phi(\Delta\text{MS}, \Delta\text{COMP})\theta + (1 - \sigma)\varepsilon \quad (8)$$

Where the first term can be broken down as the opposite of the product of three terms: (a) the product elasticity; (b) the elasticity of mark-up as described by Eq. (2) and (c) θ, the elasticity on marginal cost of automation intensity change.

Taking all those terms together, their product tends to be *negative*, so that an increase in automation AILH boosts employment. The sign of second term is not known a priori—it depends on how large capital and labor are substitutes to each other; in an aggregate Cobb-Douglas specification, the term collapses to zero; if capital and labor are strategic complements, $\sigma < 1$[13]; if there are gross substitutes the term is negative. Nevertheless, the second term increases in labor specialization ε, and declines in σ.

From the different tables, and equations above, we can further assume that κ is a decreasing function of AIM, while ϕ is a positive function of AIMS. Further, if labor is strategic complements/substitutes with other inputs, θ may be a positive/negative function of AIE. Hence, we thus clearly see that the effect of investing in automation (AILH) on employment depends on the nature of the technology, the indirect effect on product supply (and the latter is being affected by the additional mix of objectives in investing in AI).

One extreme case is when capital and labor are gross substitutes, and that the firm only uses AI for automation (AIQ = AIE = 0), and has monopoly power- in this case, labor is the only adjustment variable, leading to a decline. On another extreme, a firm invests in AI with its labor being more complement to capital and where AI leads to higher market (share) expansion, may rather lead to a net positive effect on employment. The latter is likely to be more relevant when it concerns crucial employment skills that must be bundled with new type of smart capital, as early mentioned in Brynjolfsson and Hitt (2003).

4.2 *The Econometric Labor Demand Elasticity Specification*

We now turn to an empirical specification for τ, using Eq. (8), as backbone of drivers that determine the level of τ.

Table 6 in particular demonstrates that, among firms already adopting AI, the largest segment is from companies investing both in automation and other forms of product market shifting, while the segment using AI for product market shifting is actually larger than the one doing it only for labor automation. This type of overspill to the product market is sufficiently frequent that it must controlled for in the discussion on the effect of automation on employment.

[13]The preponderance of empirical estimates on the substitution between labor and capital point to $\sigma < 1$, see Chirinko (2008).

Table 6 How firms leverage AI, % of firms

											Split sample
											Among aware AI
						Yes	19%				28%
			Yes	29%	AIM(S) > 0						
						No	10%				14%
	Yes	68%	AIL>0								
Aware of AI											
						Yes	17%				24%
			No	39%	AIM(S) > 0				No	3%	4%
	No	32%				No	25%	Pilot only			
									Yes	22%	32%

Our survey collects only data on the *direction* of employment change as a result of investing in automation. We thus consider a model of the form, I(*l*) where I = 1 if planned employment is on the decrease, 0 otherwise. As we do not have a view on output, I(I).

Using a log-linear approximation at firm level we aim to estimate a typical logistic model of the form (9):

$$I(l) = 1/1 + \exp\{\mathrm{a} + \mathrm{b}(AILS) + \mathrm{c}(AILG) + \mathrm{d}(\pi) + \mathrm{e}(\pi g) + \mathrm{f}(AIMS) + \mathrm{g}(AIM) + \mathrm{h}(\mathrm{S}) + \mathrm{cross} - \mathrm{effects} + \mathrm{fixed\ effects} + \mathrm{u} \tag{9}$$

Where the coefficients "a to h" are parameters to be estimated; *u* captures all unmeasured effects, fixed effects are industries/countries (capturing among others the common unobserved industry technical production parameters).

Remember that in our theoretical model, AILH reflects the level of maturity of investing in the variety of use cases and of AI technologies for automation. We thus build a variable as the share of AI technologies adopted by each firm for multi-use cases and for automation. This variable takes a value between 0 and 100%, as a stock effect, e.g. the amount of technologies of AI already adopted to date (denoted by *AILS*). Note as well that the equation reflects the expectations on employment changes; we thus include a flow effect, that this, the intent of a firm to invest in the future into new AI (denoted by *AILG*).[14]

Consistent with Eq. (8), we are to control for product market variables. We control for profit (π) as well as its expected change due to AI adoption, (πg). The first measure positively correlates with μ, while the second is more a summary of how firms envisage the AI diffusion pass-through into profits, as a mix of ability to expand, and/or to reap more margins. We include the fact as well that firms may choose to invest in AI to affect its product market, either via output expansion (which we denote by AIM), or via market share (AIMS).[15] All those product market effects should in fact play as interaction terms to AILS/AILG. We include them both as cross-effects and shift variables, and we let data speak. We finally control for company size (S), as it is well known that technology shape may be different by size (Dupuy and de Grip 2003).

In practice the model (9) is being estimated as a multi-logit model given three categories (increase, decrease, or no impact) of employment changes, and in difference versus industry average, so we control for trends in industry dynamics. The variables πm, πmg as well as AI*LG* are all categorical variables. Given our coding, one delta in πm is 5 points extra of current margin, one delta in πg is 3 extra points of expected margin in next 3 years, while one delta in AI*LG* amounts to 20% growth in investment in AI technologies in next 3 years. The size variable (*s*) is categorical with eight categories of employment, from 0 to 10 employees (small firms) to very

[14]We do not have any mean to split this variable in terms of investment objectives, however.

[15]We do not include AILE as otherwise, we have perfect multicollinearity.

Table 7 How firm's employment is linked to AI, direct effects

Category		Value	Sig.
Reduce needs for employment	Intercept	0.51	
	AILH	−0.023	0.101
	AIG	−0.074	0.01
	πg	−0.093	0.007
No change need for employment	Intercept	0.41	
	AILH	−0.082	0.919
	AIG	−0.031	0.001
	s (10–50 employees)	0.096	0.008

Notes: Marginal probability, default category is employment increase as a result of AIL

Table 8 How firm's employment is linked to AI, direct and indirect effects

Category		Value	Sig.
Reduce needs for employment	Intercept	0.31	
	AILH	0.24	0.05
	AIM∗AILH	−0.14	0.01
	AIMS∗AILH	−0.10	0.08
	πg∗AILH	−0.06	0.10
No change need for employment	Intercept	0.37	
	AIG	−0.04	0.01
	πg∗AILH	−0.09	0.02
	AIM∗AILH	−0.07	0.06
	AIMS∗AILH	−0.11	0.01
	Employment_10_50	0.06	0.03
	Employment_5000_10000	0.0.3	0.06

Notes: Marginal probability effect; default category is employment increase as a result of AIL

large firms (more than 10,000 employees). The largest firm size category is the default variable.[16]

4.3 Results

Our results for *total* employment are presented in Tables 7 and 8. For readability, we only reproduce statistically significant coefficients at 10% in both tables. The model always controls for industry and country effects (not reproduced). Employment increase is the reference categories, so that a negative estimated sign found in our

[16]There is a case for a selection bias in the sense that we only concentrate on firms aware of AI. However, firms not aware of AI have not given data, and if yes, noisy ones, so we can not control for them. We tried a Heckman correction where we try to predict awareness or not of AI in a first step. But it is rather difficult to have specific regressor for this first step.

model will be equivalent to a larger probability to increase (versus decrease or freeze) employment. Table 7 estimates a simple model linking employment choice to AI and profit, without the interaction terms.

In such a specification, AIM(S) do not appear significant- but we know from Table 6 that AIM positively correlates with AILH. The most significant variable is AIG (AI investment growth), then profit growth expectations, while AILH is barely significant at the margin. Nevertheless, results suggest that the more companies are vested in AILH and especially, will commit to further spend to adopt AI, the more likely they will be to increase employment, rather than reduce or not affect level of employment.

The effect of AI growth is not small; using estimated probabilities, a firm which will scale its investment budget by 20% in next 3 years, will be 65% more likely to increasing employment (18%) than the current average of 12% in the sample.

All things being equal, AI therefore *seems more employment accreditive than substitutive* for companies boosting their commitment to AI technologies. Table 8 reports results of a more appropriate specification, as variables other that AILH enter as interaction with AILH on employment changes, as per Eq. (8). Interestingly, AILH comes as a significant driver for decline in employment evolution. This effect is however counter-balanced by any plan to expand output in the form of market and market share extensions, a finding consistent e.g. with Vivarelli (2014). Likewise profit growth expectations as well as plans for further AI investment remain associated with higher employment plans.

Using the estimates, we can provide some sensitivities of AI linked to employment. Consider first a case where a firm only invested for automation-and is no longer planning to increase level of AI, while its profit growth out of AI investment is limited. In such as case, the probability to reduce (some forms of) employment is dominant (it goes to 55%) and the likelihood to increase employment collapses to zero. The opposite, and optimistic case is a firm that d continues aggressively to invest in AI (more than 20% a year), increases its profit by 3 points of margins, and further uses AI not only for automation, but for market (share) deployments.

In such as case, the probability of increasing employment becomes dominant (51%), while the probability of decline decreases to 25%, from an average in our sample of 45%. Clearly, the product market overspills are changing the distribution probabilities of employment as result of AI decision.[17]

We finally zoom by skills type in Table 9. The first column of the table uses the gross statistics shown in Table 2 to compute an indicator of net employment expectations (versus all skills' average). This indicator becomes more negative the higher the portion of firms planning to reduce, or the lower the share of companies willing to increase employment, for this skill type. Basic listening and numerical skills have 7.5 points fewer employment opportunities than the average, in relative

[17]See however the asymmetry- in this present case, employment decline has still a positive probability; while employment increase in the previous case was nil. Everything being equal, it still suggests that employment pressure may be happening along automation.

Table 9 How firm's employment is linked to AI, difference by skills

		Increase in employment linked to:		
Skills	Relative employment change (%)	AILH (%)	AIX∗AILH (%)	Total (%)
Basic listening skills	−7.50			
Basic numerical skills	−7.50			
Leadership skills	−6.50		14.4	14.4
Craft / technical skills	−5.50		16.1	16.1
Communication skills	−2.50		7.2	7.2
General management skills	−2.50		12.3	12.3
Basic IT skills	−1.50		7.30	7.3
Engineering skills	−1.50		7.6	7.6
Interpersonal skills	0.50	4.5	9.90	14.4
Optimisation and planning	0.50		10.10	10.1
Creative design skills	1.50		9.80	9.8
Project management skills	2.50		6.70	6.7
Critical thinking / problem solving	3.50	10.3	26.6	36.9
Advanced data skills	6.50	11.2	27.8	39.0
R&D skills	8.50	14.4	6.2	20.6
Advanced IT skills	11.50		9.0	9.0

contrast to advanced data or IT skills, for example, which have respectively a 6.5%/11.5% higher employability.

The other columns of Table 9 show the marginal probability from the multi-logit equation to *increasing* employment versus other categories (no change of employment or decrease). Only statistically significant coefficients at 10% are presented in Table 9. As expected, we find only rare cases where AILH is statistically associated with higher employment. However, it is remarkable that when it is, it is visible in skills with higher than average employability in the future, e.g. advanced data skills, interpersonal skills, among others. Further, an increase in AI investment growth that is aiming at increasing output boosts employment across *all* categories, except notably for skill categories with the least employability potential, e.g. *basic listening and numerical skills*. The largest potential in employment growth lies in advanced data skills, or critical thinking/problem-solving, and to a lesser extent, R&D skills, all of which are skills with relatively more employability than average. Those results are consistent with the idea that there is a tendency of skill redistribution. All those empirical findings linked to how AI influence the mix of skills are consistent with recent trends, e.g. in Deming and Kahn (2018), or Dinlersoz and Wolf (2018). What AI does seem to do is to exacerbate the trends of the recent past, we conclude.

5 Conclusions

The research above is rather new and may be extended in many ways. First, the sample can be extended and updated; second, it would be great if employment changes emerge directly for observed from data, rather than from qualitative survey. Finally, our results should be checked for robustness in terms of sample selection, in terms of omitted variables (e.g. wages as determinant of employment changes), among others.

Nevertheless, this article has put the narrative of a "workless future" to a first and new test, looking from *the derived demand side of labor by companies.* We have argued that this lens complements the recent stream of research focusing on technical automation and skills from the supply side, as corporations are primary influencers, both deciding on timing and extent of technology adoption as well as on the arbitrage to make between capital and labor, and pass-through to higher output (thus employment) or not.

Our results confirm that the narrative should indeed be more nuanced. Rather than an inevitable era of depletion of all type of jobs, our data suggest that the ultimate balance will depend on product market spillovers as well as type of skills. The product market spillover is itself dependent on how AI is used by firms- and the good news is that many firms report using AI, not (only) for pure labor automation, but for other aims, among others, expanding their product and services and competitiveness. Those are critical elements to assess how AI will be linked to employment, even if our current estimates still show an asymmetry towards lower than higher hours employment out of automation.

Regarding reallocation, our data analysis confirms a tendency towards skill-bias change. The demand for certain new skills will certainly rise, including skills linked to social, new analytics, and interfacing skills (see Deming 2017). Basic skills (including basic IT ones) exhibit lower employability and are subject to further arbitrage when companies increase their plan to invest in AI. Hence, on top of some fear of employment reduction due to automation, one may also want to ensure enough supply of skills in demand. In general, there are often frictions in the short term for new skills, e.g. STEM talents (Holtgrewe 2014 or Walvei 2016). Hence, companies in need of those new skills will have to poach the best talent in their onboarding strategy. Likewise, those companies will need to nurture their workforce via all possibilities of on-the-job and lifelong training. Perhaps this is why most of the most innovative HR practices are now coming from digital companies, from Zappos to Netflix.[18] In a world of up-skills, those companies with the right skill mix and adequate expansive business models will be those to thrive in both labor and product markets.

Acknowledgements We thank Eric Hazan and Peter Gumbel for comments and editing, as well as Dr. Chris Pissarides, Nobel Prize winner, for suggestions. All remaining mistakes are mine.

[18]https://hbr.org/2014/01/how-netflix-reinvented-hr

References

Acemoglu, D., & Resterpo, P. (2017). *Robots and jobs – Evidence from US labor markets*. NBER working paper 23285, March.

Acemoglu, D., & Restrepo, P. (2018). The race between man and machine: Implications of technology for growth, factor shares, and employment. *American Economic Review, 108*(6), 1488–1542.

Arntz, M., Gregory, T., & Zierahn, U. (2016). *The risk of automation for jobs in OECD countries: A comparative analysis*. OECD Social, Employment and Migration Working Papers, No. 189.

Atkinson, R. (2013). *Stop saying robots are destroying jobs – they are not, Technology Review*, September.

Autor, D. (2015). Why are there still so many jobs? The history and future of workplace automation. *Journal of Economic Perspectives, 29*(3), 7–30.

Autor, D., & Handel, M. (2013). Putting tasks to the test: Human capital, job tasks, and wages. *Journal of Labor Economics, 31*(2), S59–S96.

Autor, D., & Salomons, A. (2017). Does productivity growth threaten employment. Paper presented at *Cintra at the EIB meeting*, June 19.

Autor, D., & Salomons, A. (2018). Is automation labor-displacing? Productivity growth, employment, and the labor share. *Brookings Papers on Economic Activity, 2018, 1–87.*

Brynjolfsson, E., & Hitt, L. M. (2003). Computing productivity: Firm-level evidence. *Review of Economics and Statistics, 85*(4), 793–808.

Brynjolfsson, E., & McAfee, A. (2014). *The second machine age: Work, progress and prosperity in a time of brilliant technologies*. New York: W.W. Norton & Company.

Brynjolfsson, E., Rock, D., & Syverson, C. (2017). *Artificial intelligence and the modern productivity paradox: A clash of expectations and statistics*. NBER Working Paper 22401, National Bureau of Economic Research, Cambridge, MA.

Bughin, J. (2016). Reaping the benefits of big data in telecom. *Journal of Big Data, 3(1)*, 14.60(2): 338–355.

Bughin, J., & van Zeebroeck, N. (2018). Why a digital base is critical for AI. *McKinsey Quarterly*, Issue 1. https://www.mckinsey.com/business-functions/mckinsey-analytics/our-insights/artificial-intelligence-why-a-digital-base-is-critical

Bughin, J., Staun, J., Andersen, R., Schultz-Nielsen, M., Aagaard, P., & Enggaard, T. (2017). Digitally-enabled automation and artificial intelligence: Shaping the future of work in Europe's digital front-runners. *McKinsey*, October.

Chirinko, R. S. (2008). Sigma: The long and short of it. *Journal of Macroeconomics, 30*, 671–686. The CES Production Function in the Theory and Empirics of Economic Growth.

Comin, D., & Hobijn, B. (2010). An exploration of technology diffusion. *The American Economic Review, 100*(5), 2031–2059.

Deming, J. (2017). The growing importance of social skills in the labor market. *The Quarterly Journal of Economics, 132*(4), 1593–1640.

Deming, D., & Kahn, L. B. (2018). Skill requirements across firms and labor markets: Evidence from job postings for professionals. *J Labor Economics, 36*, S337–S369.

Dinlersoz, E., & Wolf, Z. (2018). *Automation, labor share, and productivity: Plant-level evidence from US manufacturing No. 18–39.*

Dupuy, A., & de Grip, A. (2003). *Do large firms have more opportunities to substitute labour than small firms*. Center for Labour Market and Social Research Working Paper.

Frey, C., & Osborne, M. (2013). *The future of employment: How susceptible are jobs to computerisation?*. Oxford Martin School Working Paper.

Garcia, A., Jaumandreu, J., & Rodriguez, C. (2002). *Innovation and jobs- evidence from manufacturing industries*. Discussion paper, Universidad Carlos III de Madrid, June.

Goos, M., Konings, J., & Vandeweyer, M. (2015). *Employment growth in Europe: The roles of innovation, local job multipliers and institutions*. Utrecht School of Economics Discussion Paper Series, Vol. 15, No. 10.

Gregory, T., Salomons, A., & Zierahn, U. (2016). *Racing with or against the machine, Evidence from Europe, ZEW*. Discussion paper 16-053.
Handel, M. (2016). *Dynamics of occupational change: Implications for the occupational requirements survey*. Research paper prepared for the US Bureau of Labor Statistics, July, bls.gov
Holtgrewe, U. (2014). New new technologies: The future and the present of work in information and communication technology. *New Technology, Work and Employment, 29*(1), 9–24.
Keynes, J.M. (1931), "The Economic Possibilities for our Grandchildren" in J.M. Keynes, *Essays in Persuasion*, Macmillan, London.
Machin, S. (2001). The changing nature of labour demand in the new economy and skill-biased technology change. *Oxford Bulletin of Economics and Statistics, 63*(s 1), 753–776.
Martinez, J. (2018) Automation, growth and factor shares. No. 736. Society for Economic Dynamics.
McKinsey Global Institute. (2017a). *A future that works: Automation, employment, and productivity*. January: *McKinsey Global Institute*.
McKinsey Global Institute. (2017b). *Artificial Intelligence – The next digital frontier?* McKinsey Global Institute, May.
McKinsey Global Institute. (2018a). *Skill shift: Automation and the future of work*. McKinsey Global Institute, May.
McKinsey Global Institute. (2018b). *Notes from the AI frontier: Modelling the impact of AI on the world economy*. McKinsey Global Institute, September.
Milkman, R., & Pullman, C. (1991). Technological change in an auto assembly plant: The impact on workers' tasks and skills. *Work and Occupations, 18*(2), 123–147.
Nubler, I. (2016), *New technologies: A jobless future or golden age of job creation?* ILO, Working paper 13, November.
OECD. (2017). *Future of work and skills*. Paper presented at the 2nd meeting of the G20 Employment Working Group, 15–17 February 2017.
Peters, B. (2004). *Employment effects of different innovation activities- microeconomic evidence*, ZEW discussion paper 4–73., Mannheim, Zentrum fur Europäische Wirtschaftsforschung.
Pissarides, Ch., & Bughin, J. (2018). *Embracing the new age of automation*, Project syndicate, January. https://www.project-syndicate.org/commentary/automation-jobs-policy-imperatives-by-christopher-pissarides-and-jacques-bughin-2018-01
Spiezia, V., & Vivarelli, M. (2000). The analysis of technological change and employment. In M. Pianta & M. Vivarelli (Eds.), *The employment impact of innovation*. London: Routledge.
Susskind, D. (2020). *A world without work: Technology, automation, and how we should respond*. London: Allen Lane.
Ugur, M., Awaworyi, S., & Solomon, E. (2016). *Technology innovation and employment in derived labour demand models: A hierarchical meta-regression analysis*. MPRA paper, Munich, paper 73557, September.
Vivarelli, M. (2014). Innovation, employment, and skills in advanced and developing countries: A survey of the economic literature. *Journal of Economic Issues, 48*, 123–154.
Walvei, U. (2016). *Digitization and structural labor market problems*. ILO research paper, 17.
Whitehead, J. C., Groothuis, P. A., & Blomquist, G. C. (1993). Testing for non-response and sample selection bias in contingent valuation: Analysis of a combination phone/mail survey. *Economics Letters, 41*(2), 215–220.
Zeira, J. (2018). Workers, machines, and economic growth. *The Quarterly Journal of Economics, 113*, 10911117.

Is Globalization Sustainable?

Luigi Paganetto and Pasquale Lucio Scandizzo

Abstract Globalization creates both positive externalities and negative effects that undermine economic growth and decrease total factor productivity and economic agents' motivation. Despite the rapid advance of digital technologies, aggregate productivity growth has slowed over the past decade or so, raising the question of how digital technologies can boost productivity. Costs and benefits are not evaluable making the real effect unclear. Nowadays, globalization seems to be necessary for the economic system, but unsustainable because of the increasing social discontent. In this context, no magic bullet is likely to exist. A crucial need is the recovery of an institutional framework where governments at all levels, are nevertheless again endowed with the function and the power to pursue wellbeing, innovation and growth through proactive economic policies.

Keywords Globalization · Sustainability · Productivity · Wellbeing

1 Introduction

The essential idea of endogenous growth is that economic development, pursued by individual agents in the name of selfish goals, creates externalities that positively impact, and in a way seemingly uninfluenced by individuals, on factor productivity. It is possible to conjecture without great effort what these externalities are. Research, education, individual economic activities all seem to be able to deploy the two characteristics required: They are undertaken for selfish reasons and provide beneficial effects that can be shared by economic agents.

Endogenous growth, on the other hand, also entails the problem of its sustainability. In the same way in which positive external effects can promote growth beyond the limits of pure individual interests, so can negative external effects

L. Paganetto (✉) · P. L. Scandizzo
FUET – Tor Vergata Economics Foundation, Rome, Italy
e-mail: luigi.paganetto@uniroma2.it; scandizzo@economia.uniroma2.it

L. Paganetto (ed.), *Capitalism, Global Change and Sustainable Development*, Springer Proceedings in Business and Economics,
https://doi.org/10.1007/978-3-030-46143-0_15

undermine the private bases of growth, depressing factor productivity and agents' motivation for productive work. The lack of convergence among the different economies can therefore also be explained by endogenous lack of development. This condition tends to be perpetuated by factors of aggravation that prevent the economy from taking off, that is to sustain in a satisfactory way a higher level of growth. Environmental externalities, such as greenhouse effects and other ecosystem changes tend to be ignored by individual countries and take the form of global threats, as in the case of climate changes. The presence of negative endogenous factors, therefore, can transform a condition of insufficient and low quality economic development into a trap, from which it may be increasingly difficult to get out. It is perhaps this trap that characterizes the present crisis of the world model of growth and the conjectures of secular stagnation that some economists have elaborated to explain the slowdown of development and transformative innovation in many countries and in the world as a whole.

2 Narratives on Globalization and Sustainability

Globalization and sustainability have become two buzzwords that are combined in different ways as part of the current economic and political discourse in two separate and contrasting narratives. As argued elsewhere (Scandizzo 2009), these narratives reflect the opposition between the modernist inclination to attribute scientific discoveries and societal transformations to unqualified social progress and the more problematic attitude of postmodernism toward the nexus between recognizable social progress and the empowerment of the elites. But it may also reflect different power positions of the parties involved, both because, as Foucault (1977, 1980) argues, science and truth are shaped by negotiating power and because discourse operates by rules of exclusion, so that power is assigned to the privileged who can speak and are listened to.

The novelty of these two contrasting views on globalization is that while the modernists spun a moderate and evolving tale of progressive improvements and ultimate success under impending great technological changes, the contrarians elaborate an extremely critical and negative counter-version, but both parties appear to agree on problematic sustainability of present development patterns. Furthermore, a shift of power seems to have occurred whereby many contrarians have been brought to power over the wings of anti-modernist narratives based on the straightforward rejection of globalism and a forceful return to new forms of nationalism variously named under the general label of "sovranism".

The two characteristics of the modernists' view of "moderate" and "evolving" optimism can be recognized in the concern for the social and environmental sustainability of the present global patterns of growth on one hand, and in the attempt to develop more creative proposals for the future. The latter include plans for more participative global governance, based on the expansion of multilateral institutions and agreements, as well as for the introduction of a new ethical capitalism, in the hope to promote radical reforms in the relationship between finance and economics and ultimately between social and private wealth.

The contrarian narrative, on its part, while enjoying a growing political success, speaks of the dark side of globalization as the loss of the soul of established local and national communities and the progressive disenfranchisement of the middle class. It also elaborates tales of increasing inequalities, a deteriorating environment under the threat of catastrophic climate change, winner-takes-all network building, and ever continuing rise of global alliances among the rich and the powerful, with multinationals taking over the world economy and keeping national governments hostages and politically powerless.

In spite of its frequent wavering and fading into seemingly preposterous tales of international plots, the contrarian narrative contains an ethical discourse that cannot be ignored by the modernists. This discourse regards globalization as a process of progressive commoditization, in which cultural traditions and social ties around the world are transformed into commercial goods bereft of sacrality or social and community value (Paganetto and Scotti 2018). In most sectors, from agriculture to ICT, commoditization is seen as determining impersonal, market driven and often dangerously de-humanizing processes of production. In agriculture, it is related to the observed alienation of smallholders as a viable social institution (family farms), the development of monocultures and loss of biodiversity. In industry, it is linked to the disenfranchisement of the urban bourgeoisie of the de-industrialized areas destroyed by the outsourcing of the global value chains. It ultimately results in the creation of massively urban biased societies, based on the unsustainable demography of the megalopolis (or the 'infinite city'). In the words of one of the most authoritative agricultural ethicists: "Agricultural producers and those who support them with technology may have been seduced into thinking that so long as they increased food availability, they were exempt from the constant process of politically negotiating and renegotiating the moral bargain that is at the foundations of the modern democratic society. Our attitude is 'full steam ahead', especially because we are expecting 3 billion additional people by 2050. The discoverers of new technologies, the gene cloners, the lawmakers who support farm subsidies, the plant breeders, the pesticide manufacturers, the organic farmers, and the globalization or protesters against agricultural biotechnology generally are unwilling to accept criticism for their actions, for all 'know' that they have made the correct choice." (Thompson 1998, p. 13).

A similar, perhaps more compelling story is told about the newly ever more pervasive digital technologies. These are told to go in a direction of a dematerialized system where not only human labor is commoditized, but it is also increasingly replaced by machines. This system is hard to predict in its details, but its prospects raise fear of a world with growing polarization of the labor market, dire prospects of unemployment, and even larger income inequalities.

This disturbing discourse is increasingly present in the political debate, even though it tends to take different forms, which may echo only some of its highlights, and often in a very blunt and easily objectionable way. The ethical aspects of science and economics have been removed from collective consciousness from the times of the positivist critique, and have been considered worthy of attention by the international science community only since environmental emergencies and climate change have become issues of global importance. But the roots of an ethical discourse for global economics and technological progress, can be recognized both in the radical critique to globalization and in the new, somewhat nervous reaction of the modernists. The underlying theme in both cases is credible in both its ethically consequentialist (consequences may be dire without appropriate standards) and proceduralist (respecting the rights of existing organisms is the foundation of our own liberty) dimensions. Instead of the economy as an activity whose main mission is to provide plenty of commodities at the lowest possible costs for all, it proposes an alternative vision based on the preservation of the environment and a careful balance between the wellbeing of present and future generations. It problematizes the relationship between man and nature in the crucial intersection with survival, access to resources and substantial freedom.

3 The Economic Problem

It may seem that the main problem related to the social effectiveness and the acceptability of globalization is its economic value. After all, cost benefit analysis is recognized as the way to proceed in the case of largest investment projects, especially those of public significance. However, it is not that simple; cost benefit ratios for the effects of globalization remain controversial because of the essentially problematic nature of the identification and measurement of the corresponding costs and benefits. In this case, as in the ethical discourse, the story told by conventional wisdom contrasts sharply with the alternative contrarian narrative. In the case of digital technologies, in particular, as recognized in a recent World Bank overarching study (World Bank 2016), while digitalization has been spreading to virtually all sectors, and firms are now more connected than ever before, productivity increases have been limited, labor markets have become more polarized, and inequality has been increasing both in developed and in developing countries. The number of democracies is apparently growing, but the share of free and fair elections is falling.

The World Bank Report (2020) on Trading for development in the age of global value chains showed that goods-producing value chains have become less trade-intensive, the cross-border services are growing more than 60% faster than trade in global value and chains are becoming more knowledge-intensive and reliant on high-skill labor. Finally, global value chains are being reshaped by cross-border data flows and new technologies, including digital platforms, the Internet of Things, automation and Artificial intelligence. These trends we identify may favor advanced economies, given their strengths in innovation and services as well as their highly skilled workforces. But the challenges are getting steeper for countries that missed the last wave of globalization. In the case of Italy (Zingales and Pellegrino 2017) there is no evidence that the lack of growth in labor productivity is due to trade dynamics, Italy's inefficient governmental apparatus, or excessively protective labor regulations. By contrast, the data suggest that Italy's slowdown was more likely caused by the failure of its firms to take full advantage of the ICT revolution.

Quantitative studies of the effects of globalization have been conducted mainly by the advocates of the benevolent interpretation. Their ultimate claim is that undesirable developments in labor markets and income distribution and productivity gains are happening not because of globalized technological advances, but in spite of them. "Dividends" from digitalization, for example are estimated as potentially overtaking any corresponding costs and adverse effects, as the digital divide is closed and access to the network is assured to a larger proportion of the 60% of the world population that is still excluded from it. In monetary terms these dividends can be quantified as increases in incomes or consumer surpluses consequent on the application of the technology whose discovery and development have been facilitated by the past 50 years of research and by its still initial application.[1]

Costs include direct research and development (R&D) costs (and some imputed costs of permanent installations), and education and training costs. Costs related to the training of researchers and the adverse consequences for other agents are typically ignored. Also, market rather than shadow prices, and partial equilibrium analyses are generally used and the counterfactual situation (i.e. the situation that would have obtained had the research not been conducted) is essentially identified with the status quo. Finally, neither the risks involved, the opportunities foregone nor the irreversible nature of many of the resources committed to research and the subsequent developments are considered. In addition, there are doubts about the orders of magnitude of the economic returns reported in the literature deriving from the widely different methodologies, the confusion between nominal and real returns, the varying time lags, the differences between ex ante and ex post rates and the systematic downward estimates from self-evaluations.

Furthermore, rather than being based on average (or median) benefits and costs, modern economic evaluation should be based on an estimate of the impact of agricultural research on the contingent wealth of the winners and losers. In particular, such investment may destroy and create real options, i.e. a combination of capabilities and exposure to opportunities and risks, whose economic value may go well beyond, both positively and negatively, the estimated aggregate income (or consumer surplus) increases in average scenarios. These options include reduced or enhanced adaptability to climate change—a key factor for economic performance and perhaps for survival in the years to come.

[1]R. Baldwin (2019) believes that Globalization 1.0 and 2.0 favored the G7 countries, while the new globalization 3.0 is definitely favoring countries such as India and China. The reason, in his opinion, is that after a globalization that brought about the shift of goods and services but left investments in innovation in the G7 countries, today the ICT technologies allow the transfer of knowledge according to the axis of the Global value chain. The explosive development of digital technologies creates the possibility of creating artificial intelligence, which in turn may favor emerging countries that can adjust labor skills at much lower costs than in the G7 countries.

4 The Tradeoff Between Freedom and Wellbeing

In order to characterize the disruptive challenge of globalization to the traditional social contract, we must start from the concept of freedom. A popular, if elementary way to define this concept is in terms of lack from interference: in the absence of specific provisions, each citizen has the right to carry out all actions that she deems appropriate, unless these actions interfere, with citizens' freedom and, by threatening a similar right, on the actions that can be carried out by somebody else. This very general definition is linked to the core problem of the establishment of rights in modern societies. As John Rawls (1971) perceptively conjectured, about justice: if someone were asked to choose the basic rules by which society will be run, but without knowing anything about her personal circumstances—how much intelligence, talents or wealth, then from behind this "veil of ignorance" two key principles would likely to be chosen by a rational person: the *liberty principle* and the *difference principle.* The *liberty principle* holds that people would require minimal civil and legal rights to protect themselves, while the *difference principle* states that everyone would want to live in a society where improving the condition of the poorest is an absolute priority, since, given the "veil", one would want to minimize the risk of falling into that category.

In the Rawlsian conception, rights constitute at the same time contingent assets and liabilities, since a right is an "entitlement or justified claim on others". These require both negative and positive obligations, meaning that others are required to not harm anyone (negative obligation) and surrender portions of their earnings for the benefit of low-income earners (positive). The social contract is thus a network of weak and strong ties, where enforcement and incentives are balanced by the dominance of "local" obligations over "global" rights, which arise necessarily from the rational choice, under the veil of uncertainty, to commit resources to the less fortunate members of society.

While market integration and internationalism was for long time reflecting progressive social views, of which John Rawls was the most eminent philosophical exponent, at the end of the 70s a variety of historical and political circumstances provided the basis for the alternative views to prevail from another philosophical school to prevail. For Nozick, a libertarian exponent who expanded his views in a 1974 book, graphically called "Anarchy, State and Utopia", nothing beyond the most "minimal" non-voluntary obligations to other members of society can ever be justified, and weak ties dominate strong ones in a social universe where globalization is made inevitable by its minimalist social requirements and its rational and progressive features. For this school of thought, Rawls' theory of justice impossibly denies the possibility of mutually advantageous transactions, in the sense that the outcome of the economic process aims to a state where "at the end" everyone is equal. But, if only the end result matters, anybody who can be dispossessed to ensure equality holds (i.e. everybody is under the threat of expropriation).

Instead of an end-state theory, globalization has progressively embraced Nozick's historical theory, which claims that social justice depends not only on some general notion of "what is right", but on the more specific question of what kind of entitlements are justified in the light of today's entitlement structure and the past history of transfers and acquisitions. These are completely justified if they are the fruit of voluntary and mutually beneficial exchanges and holdings are initially acquired in a just way.

Global adoption of free market liberalism and the more radical Nozick's principles has characterized the push to globalization that has disrupted and re-organized the international economic order since the late 70s. A great acceleration of the process of international trade and finance has been one main phenomenon of this drive, accompanied by structural changes such as the integration of the value chains across countries and the great expansion of digital connectivity and associated technologies. A major effect has been the growing divergence in the practice of Rawls' *liberty* and *difference* principles, with spectacular successes of economic and technological progress, combined with rampant social injustice affecting not only the poor but also the middle class. Free market liberalism in fact is based on the belief that people require fundamental autonomy and a just society should only be committed to protect their (negative) rights not to be disturbed by other individuals, who are allowed to take action for their own benefit. For example, the right to pursue happiness is the freedom to take those actions that the individual sees fit to better her life, so long as they do not infringe upon the rights of others. This implies that the role of the state is not one of proactively seek social justice, through appropriate policies, but it is basically of regulatory nature. It consists, in other words, in protecting the individual's right to act and in preventing these actions from negatively interfering on each other's rights. In this view, while the government agenda is minimal, the role postulated is still loaded with the danger that, in interposing its actions to "protect and prevent", it may negatively infringe in individuals' basic rights. The very existence of the government, in other words, indispensable as it may be, appears a threat on private rights for the citizens of any society.

By espousing free market liberalism and, in many ways, the more extreme version proposed by Nozick, globalization has created a world community increasingly based on the negative rights of not being interfered in the exercise of one's freedom, combined with the right to pursue any action that does not interfere with others' rights and is not explicitly recognized as illegitimate. This latter form of right is an active one, since it requires the possibly aggressive pursuit of one's goals. The positive side of freedom, on the other hand, has been progressively neglected. The rights to expect from others, and the state, some fundamentals that go beyond the simple commitment to abstain from negative interference have been essentially abandoned as governments and multilateral institutions have been confined to the role of impartial "guardians of the markets" in spite of rampant processes of impoverishment and disenfranchisement of larger and larger parts of the population.

5 Conclusions

In many ways, globalization seems to have resulted in a situation with no apparent way out, where the world cannot give up globalization and yet this does not seem ultimately sustainable because of the increasing social discontent. This "trap" is made more dramatic by two related facts: (1) the promises of increased productivity linked to the new "global" technologies, ICT, robotic, artificial intelligence have not yet been fully realized, (2) social problems such as unemployment, income distribution and relative poverty are increasing while economic growth is declining.

Digital transformation represents an opportunity for improving productivity growth by enabling innovation and reducing the costs of a range of business processes. Yet despite the rapid advance of digital technologies, aggregate productivity growth has slowed over the past decade or so, raising the question of how digital technologies can boost productivity work shows there is hope for the future. What is sure is that all countries are committed to increasing productivity through innovations and making the effort necessary to appropriate them internationally. The current US-Chinese dispute is presented as a war of duties but can also see as a controversy over technological leadership and innovation as a central aspect of the ongoing globalization phenomena. In this context it is decisive how the Institutions (G20, WB, WTO, IMF) act in the global Governance. What role does the EU have in this framework? It should not be forgotten that one of EU main weaknesses is (the lack of) a coherent foreign policy.

As to the sustainability of globalization processes, the contrarian narrative appears more plausible to most: the problem is to what extent a seemingly failing global governance is able to control conflicts through adequate policies that could end up being explosive. To what extent globalization can be reconciled with social consensus and social expectations without a sufficient and growing amount of innovation, productivity growth and income re-distribution? Here the contrarian and pessimistic narrative seems to prevail, in line with the broad dissatisfaction and distress of the working middle class in deindustrialized developed economies. The "new" labor policies, austerity and minimal government, combined with the inability to tackle economic crises, have all been instrumental in determining an environment, where job insecurity, youth unemployment, and economic and political risks are rampant, public capital is declining, the environment is deteriorating and governments are increasingly powerless and confined to the sidelines as passive regulators of often failing markets. Population movements in time and space play an important but highly uncertain role in a phase of unclear demographic transition, where migration appears both an uncertain promise to the poorest countries and a dire threat to the richer ones. The declining wellbeing of the middle class, a designated victim of future labor saving technologies, appears to worsen social injustice and further instability and income inequality.

Against this background of chaotic and ungoverned economies and population movements, a new concept of economic governance is advancing. This concept requires prioritizing the ecosystem services of the different components that contribute to land use (forests, rural areas, watersheds, urbanized vast areas, etc.) by adopting a set of values or shadow prices that make the land use scale hierarchical and compatible with the functionality of potential ecological networks. This new form of governance is also combined with an innovative idea of a holistic infrastructure encompassing social disruption and complex spatial realities such as growing urban areas and changing landscapes. According with this concept, modern public capital results from the integration, at a territorial scale, of a series of components, including traditional elements of produced capital such as roads and railways, combined with assets and services from human capital and the ecosystem. These "new" components include social settlements, employment patterns, natural resources, biodiversity and various forms of tangible and intangible physical and cultural services such as water provision, carbon capture and landscapes. They are new not in the sense that they were not there before, but that a new form of awareness of their vital function has projected them from the background to the forefront of modern infrastructure as a set of complex and integrated spatially built systems (Damania et al. 2019).

Though only very tentative, these new economic concepts appear to suggest a way to the contradiction between a pattern of development which seems to deny both sustainable economic growth and feasible social justice. While no magic bullet is likely to exist, a crucial need is the recovery of an institutional framework where governments at all levels, without discarding their essential regulatory functions, are nevertheless again endowed with the function and the power to pursue wellbeing, innovation and growth through proactive economic policies. These policies should go hand in hand with a major role of multilateral institutions and a new pact on the international economic order, giving priority to measures aiming to create productivity increases and, at the same time, social consensus and wellbeing.

References

Baldwin, R. (2019). *The great convergence: Information technology and the new globalization.* Belknap Press: An Imprint of Harvard University Press, Reprint edition.

Damania, R., Desbureaux, S., Rodella, A., Russ, J., & Zaveri, E. (2019). *Quality unknown: The invisible water crisis.* Washington, DC: World Bank.

Foucault, M. (1977). *Discipline and punish.* New York: Pantheon.

Foucault, M. (1980). Two lectures. In C. Gordon (Ed.), *Power/knowledge: Selected interviews.* New York: Pantheon.

Nozick, R. (1974). *Anarchy, state and utopia.* New York: Basic Book. ISBN: 9780465097203.

Paganetto, L., & Scotti, V. (2018). Globalization, inclusion and sustainability in a global century. In *Globalization, inclusion and sustainability in a global century.* Rome: Eurilink University Press.

Rawls, J. (1971). *A theory of justice.* Cambridge, MA: Belknap Press of Harvard University Press.

Scandizzo, P. L. (2009). Science and technology in world agriculture: Narratives and discourses. *AgBioForum, 12*(1), 23–33. Available on the World Wide Web: http://www.agbioforum.org

Thompson, P. B. (1998). *Agricultural ethics*. Ames: Iowa State University Press.
World Development Report. (2016). *Digital dividends*. World Bank Group.
World Development Report. (2020). *Trading for development in the age of global value chains*. World Bank Group.
Zingales, L., Pellegrino, B. (2017). *Diagnosing the Italian disease*.

The manufacturer's authorised representative in the EU is Springer Nature Customer Service Centre GmbH, Europaplatz 3, 69115 Heidelberg, Germany. If you have any concerns regarding our products, please contact ProductSafety@springernature.com

Printed and bound by CPI Group (UK) Ltd, Croydon, CR0 4YY
22/09/2026
02221060-0009